Three-dimensional nets and polyhedra

Wiley monographs in crystallography

Polymorphism and polytypism in crystals
Ajit Ram Verma and P. Krishna

Molecular crystals
J. L. Amoros and M. Amoros

Color and symmetry
Arthur L. Loeb

Fourier methods in crystallography
G. N. Ramachandran and R. Srinivasan

Three-dimensional nets and polyhedra
A. F. Wells

Three-dimensional nets and polyhedra

A. F. Wells

Department of Chemistry
and
Institute of Materials Science
University of Connecticut

A Wiley-Interscience Publication

JOHN WILEY & SONS, New York • London • Sydney • Toronto

Library of Congress Cataloging in Publication Data:

Wells, Alexander Frank.
Three-dimensional nets and polyhedra.

(Wiley monographs in crystallography)
"A Wiley-Interscience publication."
Includes bibliographical references and index.
1. Crystallography, Mathematical. I. Title.
QD911.W43 548′.7 76-49089
ISBN 0-471-02151-2

Printed in the United States of America

10 9 8 7 6 5 4 3 2 1

Foreword to the series

In the 1920's comparatively few people paid much attention to the branch of science known as crystallography, and the only journal devoted to it was circulated to a few hundred subscribers, mostly libraries. Scientists in the classical disciplines hardly recognized crystallography as a science, although each apparently regarded it as a small segment of his own field. No American university boasted a professor of crystallography and any instruction given was adjunct to mineralogy. Nevertheless, papers of interest to crystallographers appeared in journals of mineralogy, physics and chemistry, although not in large numbers. In those days one might claim to have an all-around acquaintance with crystallography because he was able to keep up with the literature.

This is no longer true. In the period between the first and second world wars, science flourished, and scientists not only published more papers, but an increasing proportion of them dealt with solid materials. It was inevitable that the chemists, metallurgists, physicists and ceramists should make increasing use of crystallographic theory and methods, that the journals of many fields should publish more papers of crystallographic interest, and that new journals devoted to crystallography and the "solid state" should arise. Soon the abstracting journals contained hundreds of titles of crystallographic interest with each issue, and now few of us can keep even reasonably well informed about the many aspects of the science of crystals, to say nothing of keeping abreast of the advances in all these aspects. Not only is it out of the question to keep up with the mass of literature that is turned out, but it is even a little difficult to maintain contact with all the advances in one's own specialty. Accordingly, we are tending to become parochial.

In the words of Warren Weaver ". . .the volume of the appreciated but not understood keeps getting larger and larger." In order to improve this condition to some extent we need the services of those who, having become authorities in some segments of our field, are willing to integrate their understandings of these limited regions. With such help many of us can gain a sufficient understanding of matters whose original literature we have neither the time nor the inclination to study. Such writings exist in several fields, but none, to date, in crystallography. It is to fill this need that the Wiley Monographs in Crystallography are offered.

MARTIN J. BUERGER

Preface

The following account represents an attempt to summarize the results of studies of three-dimensional systems of linked points, a subject that has interested the author during the past 20 years. Some of the results have been published as a series of papers in *Acta Crystallographica* entitled "The Geometrical Basis of Crystal Chemistry," Parts 1 to 12, 1954–1976. * During the past few years many new 3D nets and some new 3D polyhedra have been derived, but it has become obvious that the publication of further papers in this series is not a satisfactory way of describing the work. So many cross-references to earlier papers became necessary that the later papers tended to be unintelligible unless the reader had all the previous papers readily available and also had the patience to follow the sometimes devious lines of thought of the author. This book summarizes the earlier work and includes descriptions and illustrations of many new systems that have not been described.

The obvious and simple relation of much of this work to the Platonic solids and to simple Euclidean geometry makes it surprising that geometers have not explored this field during the past two thousand years or so. The reason is presumably that the study of *periodic* three-dimensional systems of points, lines, and volumes, seems to have been left for the most part to the crystallographer, despite notable contributions from a small number of mathematicians interested in three-dimensional geometry.

Apart from their intrinsic interest, and beauty, as examples of the logical extension of classical Euclidean geometry, three-dimensional systems of connected points are clearly very much a part of structural chemistry in its widest sense. The relation of the crystal structures of compounds such as zeolites and clathrate hydrates to space-filling arrangements of polyhedra is evident and has long been recognized; it is also well known that many silicates and aluminosilicates have structures based on vertex-sharing tetrahedra placed at the points of various 4-connected nets. The relation of 3-connected [and (3, 4)-connected] nets to a (smaller) number of crystal structures has

* "The Geometrical Basis of Crystal Chemistry," Parts 1–12, *Acta Crystallogr.*, pt. 1, 1954, **7**, 535; pt. 2, 1954, **7**, 545; pt. 3, 1954, **7**, 842; pt. 4, 1954, **7**, 849; pt. 5, 1955, **8**, 32; pt. 6, 1956, **9**, 23; pt. 7 (with R. R. Sharpe); 1963, **16**, 857; pt. 8, 1965, **18**, 894; pt. 9, 1968, **B24**, 50; pt. 10, 1969, **B25**, 1711; pt. 11, 1972, **B28**, 711, pt. 12, 1976, **B32**, 2619.

become apparent more recently. It is still unusual, however, for the crystallographer to describe a structure in terms of its basic topology. Such a description not only provides a simple and elegant way of representing the structure but it emphasizes relations between structures that are not always apparent from conventional descriptions in terms of space groups and sets of equivalent positions.

It has not been easy to write a connected account of this work, and some of the text may be difficult to follow when the reader is assisted only by line drawings and pairs of stereoscopic photographs. In this subject models are indispensable, and the author has had the advantage of studying models of most of the systems described. Models of many of the 3D polyhedra can be constructed quite easily from strips of thin card or plastic, and many 3D nets may be built from the plastic tubing and metal "valence clusters" that are commercially available.

A. F. Wells

Storrs, Connecticut
September 1976

Acknowledgments

I acknowledge the contribution to the earlier work on 3D polyhedra of R. R. Sharpe (who also constructed some of the models of 3-connected nets) and the help received from my colleagues Profs. M. J. Buerger, B. L. Chamberland, and R. Schor at the University of Connecticut. I also wish to express my gratitude to the University of Connecticut Research Foundation for a grant that together with much practical assistance from Dr. J. Haberfeld made possible the stereoscopic photography of many new models.

A. F. W.

Contents

Three-dimensional nets and polyhedra

PART I

Three-dimensional nets

1

Introductory

The present study began by attempting to answer two questions. Only five simple convex polyhedra (n, p) have all their faces of the same kind (n-gons) and the same number (p) meeting at each vertex:

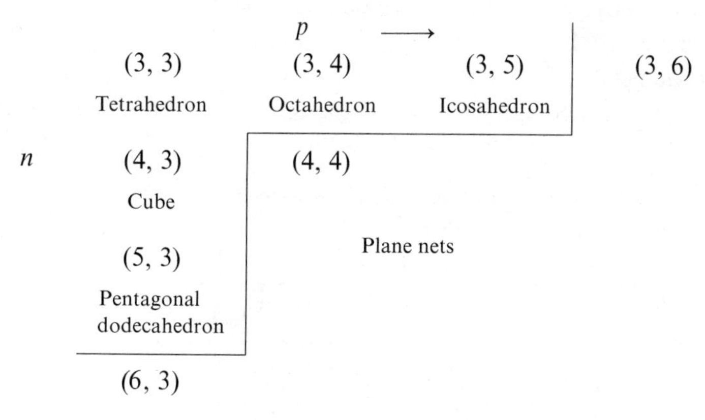

The tetrahedron is the first member of the family (3, 3), (4, 3), and (5, 3); in all members three edges meet at each vertex (3-connected), and the shortest circuits (n-gons) have 3, 4, or 5 edges. The next member of this family is the plane hexagonal net (6, 3), which has the property that the shortest circuit including any two of the three links meeting at any point is a 6-gon. Our first question concerns the nature of 3-connected nets $(n, 3)$ in which n is greater than 6. Such nets, in which the shortest circuit including any pair of links from any point is an n-gon, are called *uniform* 3-*connected nets*; evidently they contain no circuits smaller than n-gons.

Alternatively we may focus our attention on the circuits (polygons) that form the surface of the polyhedron. All the faces of three of these polyhedra, tetrahedron, octahedron, and icosahedron, are triangles (which are equilateral in the most regular forms of the solids), and the numbers of triangles meeting at each vertex are respectively 3, 4, and 5. If six triangles meet at each point, a closed convex polyhedron is no longer possible, and (3, 6) is a tessellation on an infinite two-dimensional (2D) surface and is the regular plane net (3, 6) on the Euclidean plane if the triangles are equilateral, or on an infinite (open-ended) cylindrical surface. If now we make a tessellation in which seven or more triangles meet at each point and make it from narrow strips of paper or plastic, which in the simplest case are made equal in length, we find that the surface buckles and eventually joins up in a complex way. Our second question concerns the nature of the surfaces on which tessellations (3, p) can be drawn when p exceeds the value (6) for a plane net. We refer to such surfaces as the surfaces of 3*D polyhedra.*

These two families, 3-connected nets (n, 3) and 3D polyhedra (3, p), form only part of the whole problem (Table 1.1), which includes all systems (n, p) having n and/or p greater than the values permissible for plane nets. Since we derive separately the uniform 3D nets (and also some other nets which are not uniform nets) and 3D polyhedra, and since systems of both types appear in Table 1.1, it is necessary to discuss the relation between them. A closely

Table 1.1 Systems (n, p) of connected points

	p						
n	3	4	5	6	7	8	...
3	t	o	i	(3, 6)			
4	c	(4, 4)					
5	d						
6	(6, 3)			3D Polyhedra and 3D nets			
7							
8							
⋮							

related question is the dual relation between pairs such as (3, 8) and (8, 3). However it is not feasible to consider such basic questions until we have described some 3D nets and polyhedra, and discussion of these matters is therefore deferred.

There are only five convex polyhedra having all faces of the same kind and the same number of faces meeting at each vertex, and there are only three plane nets having all polygons of the same kind and the same number of polygons meeting at each point. Moreover, there are two ways of accounting for these circumstances. The *topological* proof is concerned only with the number (n) of edges of the faces or polygons and with the connectedness (p) of the vertices (points), that is, the number of edges (links) meeting at each vertex (point). For a (finite) convex polyhedron or a plane net, this number p is the same as the number of n-gons meeting at a point, since two edges of each polygon meet at a given point and each edge is common to two polygons. For a finite convex polyhedron (n, p) with Z vertices, it follows directly from Euler's relation (which itself is purely topological) that

$$Z = \frac{4n}{4 - (n - 2)(p - 2)} \qquad (1.1)$$

For finite values of Z,

$$(n - 2)(p - 2) < 4 \qquad (1.2)$$

which is satisfied by only five integral combinations of n and p, namely:

(n, p)	Z	
(3, 3)	4	tetrahedron
(3, 4)	6	octahedron
(3, 5)	20	icosahedron
(4, 3)	8	hexahedron
(5, 3)	12	dodecahedron

The solutions for $Z = \infty$, that is, of

$$(n - 2)(p - 2) = 4 \qquad (1.3)$$

correspond to the plane nets (n, p), that is, (3, 6), (4, 4), and (6, 3) (Fig. 1.1*a*).

Equation (1.1) also gives Z for the Archimedean and Catalan (semiregular) solids if the *mean* value of n or p is used. The 13 Archimedean solids have

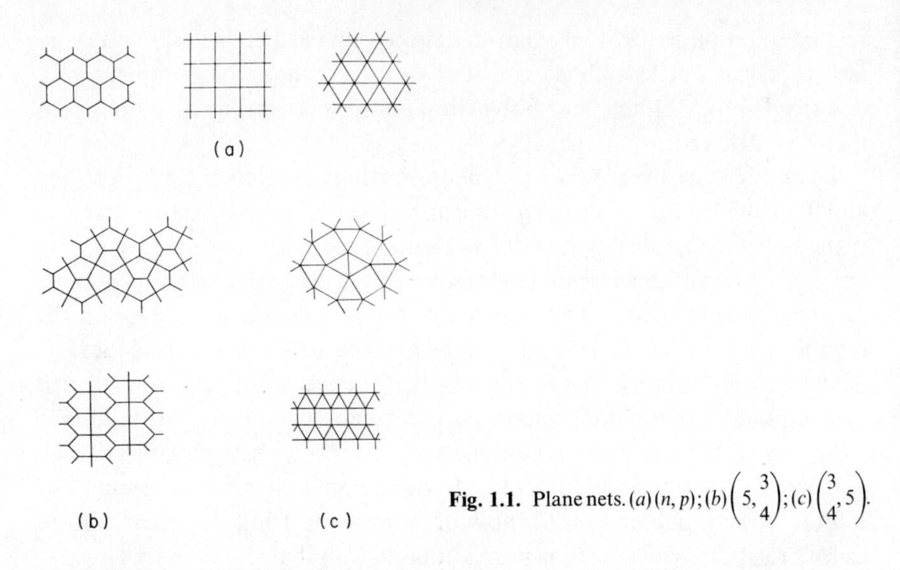

Fig. 1.1. Plane nets. (*a*) (n, p); (*b*) $\left(5, {3 \atop 4}\right)$; (*c*) $\left({3 \atop 4}, 5\right)$.

$p = 3, 4,$ or 5 (the same values as the Platonic solids) but with n ranging from 3 to 10 and nonintegral values of n_{mean}, therefore also of the product $(n - 2)(p - 2)$. The Catalan solids are their duals, with nonintegral values of p_{mean}, for example:

Z	n	p		
12	$\frac{24}{7}$	4	cuboctahedron	$(n - 2)(p - 2) = \frac{20}{7}$
14	4	$\frac{24}{7}$	rhombic dodecahedron	

The numbers of solutions of (1.2) and (1.3) are not dependent on the geometry of the systems; that is, they do not require that the n-gons be regular (equilateral and equiangular) or that the edges be straight lines. It is important to distinguish between the terms equilateral and equiangular. If a plane triangle is equiangular it is also equilateral; that is, regular; if $n > 3$ an n-gon may be equiangular but not equilateral (e.g., rectangle), or it may be equilateral but not equiangular (e.g., rhombus). Thus a plane equilateral triangle is equiangular (angles 60°), but on the surfaces of varying curvature that we meet later, a triangle may be equilateral but not equiangular.

The *geometrical* proof of the existence of only five *regular* solids and of three *regular* plane nets makes use of the fact that the internal angle of a regular n-gon is equal to $2(n - 2)/n$ right angles. For a convex polyhedron the sum of the p angles meeting at a vertex must be less than 360°, and we write

$$\frac{2p}{n}(n - 2) < 4$$

which can be rearranged to $(n - 2)(p - 2) < 4$. For a regular plane net the condition is $(n - 2)(p - 2) = 4$. This geometrical approach, which assumes regular plane polygons, thus gives the same result for polyhedra and plane nets as the purely topological one. If the meaning of n is extended to include the mean value, the solutions for plane nets include all plane 3- and 4-connected nets. The "missing" solutions, for $n = 5$ and the dual with $p = 5$, imply nonintegral values of p or n, respectively. The two solutions for $n = 5$ and their duals are in Fig. 1.1*b* and *c*.

It is convenient to use the term *Platonic solid* to mean the topological entity (n, p) and to keep the term *regular solid* for the configuration of (n, p) having regular plane faces. Similarly we shall apply the terms Archimedean and Catalan to polyhedra with the topological characteristics of these two groups of polyhedra without implying the geometrical properties they possess as semiregular solids. The same terms also are applied to 2D and 3D nets of the types (n, p), $\left(\frac{m}{n}, p\right)$, and $\left(n, \frac{p}{q}\right)$, respectively.

Our interest lies largely in *periodic* systems of connected points, but for the sake of completeness we note that there are other solutions for systems such as $(3, p)$ for $p \geqslant 7$. They include radiating nets, on which we comment briefly later, and hyperbolic tessellations. The three triangulated regular solids may be inscribed on a sphere when the internal angles of the triangular faces are 120°, 90°, and 72°, respectively, the edges being curved, convex outward. In a tessellation of triangles on this surface of constant positive curvature, a maximum of five can meet at each point if all the points are to be topologically equivalent. If the sphere is opened out to become a Euclidean plane, a surface of zero curvature, a maximum of six triangles can meet at each point of a tessellation of equivalent points. For an equilateral triangle on the Euclidean plane, the internal angle is 60°. It is possible to imagine a surface of constant negative curvature on which equilateral triangles having internal angles less than 60° could be inscribed with $p \geqslant 7$. Such tessellations have been described,[1] but since the surface is not realizable in ordinary space, they do not concern us here.

Since the solids and plane nets (n, p) correspond to integral values of $(n-2)(p-2)$ from 1 to 4 (and the Archimedean and Catalan solids to nonintegral values),

$(n-2)(p-2)$	n	p	
1	3	3	tetrahedron
2	3	4	octahedron
	4	3	hexahedron
3	3	5	icosahedron
	5	3	dodecahedron
4	3	6	plane nets
	4	4	plane nets
	6	3	plane nets

we may set out these (n, p) systems according to the values of $(n-2)(p-2)$ (Table 1.2). An important feature of Table 1.2 is that whereas each entry to

Table 1.2 Values of $(n-2)(p-2)$ for systems (n, p)

n \ p	3	4	5	6	7	8	9	10	11	12
3	1	2	3	④	5	6	7	[8]	9	10
4	2	④	6	[8]						
5	3	6	9	12						
6	④	[8]	12	16						
7	5									
8	6									
9	7									
10	[8]									
11	9									
12	10									

④ Plane nets

[8] 3D nets of Table 1.3

the left of the heavy line corresponds to a single entity (polyhedron or plane net), an entry in the remainder of the table *may* correspond to more than one 3D system (n, p). In some cases, for example (12, 3), it happens that only one net is at present known, but in other instances the same symbol (n, p) represents a number of topologically different systems [e.g., 15 different (8, 3) nets]. Therefore we shall consider in Chapter 3 other topological properties of nets that distinguish between some at least of a number of nets having the same symbol (n, p). Even if we restrict ourselves to systems (n, p) having a configuration in which all the links are of equal length and all the interbond angles are equal, we find in many cases more than one (n, p). For polyhedra and plane nets the topological derivation gives the same number of solutions (n, p), as does the geometrical derivation assuming regular polygons; but this is not true for the 3D systems because there is no unique configuration of a nonplanar n-gon even if it is both equilateral and equiangular.

It appears that the integral solutions of

$$(n - 2)(p - 2) = 8 \tag{1.4}$$

include some simple and important 3D nets. If we again include the solutions for the "missing" values of n and p, which imply nonintegral values of either p or n, we arrive at the values given in Table 1.3. The entries in the left-hand half include the three regular plane nets in the first column and also the eight Archimedean and eight Catalan semiregular plane nets, if the value of either n or p is taken as the *mean* value. The net (3, 10) corresponds to a surface

Table 1.3 Nets corresponding to solutions of $(n - 2)(p - 2) = 4$ or 8

2D nets $(n-2)(p-2)=4$				3D nets $(n-2)(p-2)=8$			
n and p both integral		n or p nonintegral		n and p both integral		n or p nonintegral	
n	p	n	p	n	p	n	p
3	6			3	10	$\frac{22}{7}$	9
		$\frac{10}{3}$	5			$\frac{10}{3}$	8
4	4			4	6	$\frac{18}{5}$	7
		5	$\frac{10}{3}$			$\frac{14}{3}$	5
6	3			6	4	5	$\frac{14}{3}$
		(Catalan if $p = p_{\text{mean}}$)				7	$\frac{18}{5}$
(Archimedean if $n = n_{\text{mean}}$)				10	3	8	$\frac{10}{3}$
						9	$\frac{22}{7}$

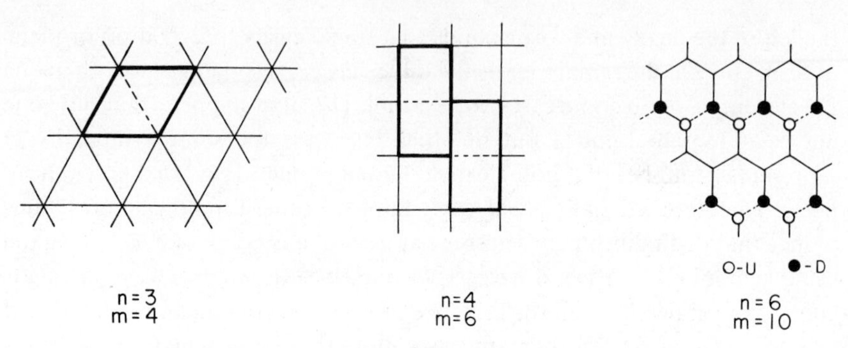

Fig. 1.2. The next larger circuits in the plane nets (n, p).

tessellation (or 3D polyhedron). The appearance of 4-, 6-, and 10-gons in 3D nets corresponding to integral solutions of (1.4) is to be expected for the following reason. (For a different approach see Chapter 4.) These polygons (m-gons) are the next larger circuits in the plane nets (n, p), where $m = 2n - 2$, n being 3, 4, or 6 (Fig. 1.2). Very simple 3D nets can be constructed by omitting certain of the links in these plane nets and joining the $(p - 1)$-connected points so created to similar points in nets placed above and below those of Fig. 1.2. For example, by joining the points marked D in one layer to points U of a similar layer below, a 3D(10, 3) net is formed. Since $m = 2n - 2$, $(m - 2) = 2(n - 2)$; thus $(m - 2)(p - 2) = 2(n - 2)(p - 2) = 8$, where m and p are the values for the 3D net.

Examples will be given of all the 3D nets enclosed within the heavy lines in the right-hand portion of Table 1.3; systems with nonintegral values of n_{mean} have not been studied. Of course, the systems in Table 1.3 with p non-integral are not included in Table 1.2, which does include some nets that do not appear in Table 1.3—for example, (7, 3), (8, 3), and (9, 3) and their duals, for which $(n - 2)(p - 2) = 5$, 6, or 7, respectively, and also (12, 3) and (3, 12), for which $(n - 2)(p - 2) = 10$.

As noted earlier, we postpone a consideration of the relation between 3D polyhedra and 3D nets until the final section of the book. Since an understanding of 3D polyhedra depends on a knowledge of the basic 3D nets, we deal first with nets, then with 3D polyhedra. We group nets according to the value of p, describing first the nets (12, 3), (10, 3), (9, 3), (8, 3), and (7, 3) in which all points are 3-connected, then nets containing both 3- and 4-connected points (which include the nets $(7, \frac{18}{5})$, $(8, \frac{10}{3})$, and $(9, \frac{22}{7})$ of Table 1.3), and finally 4-connected nets. Reference is made to some more highly connected nets ($p > 4$), but these have not been studied systematically. The net $(5, \frac{14}{3})$ of Table 1.3, which corresponds to the pyrite (FeS_2) structure, is described in Chapter 3. Some properties of these nets are listed in Table 3.2.

References

1. H. S. M. Coxeter and W. O. J. Moser, *Generators and Relations for Discrete Groups*, Berlin: Springer, (1957), p. 53.

2

Uniform nets

We now examine the significance of the symbol (n, p) when applied to a three-dimensional net. If plane nets are regarded as ways of dividing the plane into polygons, the analogous problem in three dimensions would appear to be the partitioning of space into polyhedral compartments, a subject that has received some attention. von Fedorov listed the five polyhedra that fill space when similarly oriented, Andreini described the much more numerous space-filling combinations of regular and Archimedean semiregular solids, whereas Coxeter's three skew polyhedra are half-space-fillings by cubes, truncated octahedra, and equal numbers of tetrahedra and truncated tetrahedra.[1–3]

Alternatively the Platonic solids and plane nets (n, p) may be regarded as ways of connecting points so that each is connected to the same number of others to form systems in which all the shortest circuits are n-gons. The analogous problem in three dimensions is then to enumerate networks of points that repeat regularly in space and have the property that the shortest circuit, starting from any point and including *any two* of the links meeting at the point, is an n-gon. Such a net may be described as a *uniform net*. Note that this definition does not require that p be the same for all points. If every point is 3-connected, such uniform nets represent the continuation into three dimensions of the series starting with three of the Platonic solids and the only $(n, 3)$ plane net:

(n, p)	:	(3, 3)	(4, 3)	(5, 3)	:	(6, 3)	:	(7, 3)	(8, 3)	(9, 3)	(10, 3)	(12, 3)
		Platonic solids				Plane net		Uniform 3D 3-connected nets				

The problem can also be studied for $p = 4$, 5, and so on.

We shall also expect to find 3D nets corresponding to Archimedean and Catalan solids, giving the following classes:

Uniform (n, p) nets	Platonic	all shortest circuits n-gons; all points p-connected
	Catalan	all shortest circuits n-gons; points p-, q-, and so on, connected
	Archimedean	shortest circuits of more than one kind; all points p-connected

As examples of Catalan-type nets we describe several nets containing both 3- and 4-connected points (Chapter 8) and also the "pyrite" net, which contains 4- and 6-connected points. Archimedean-type 3D nets are numerous, and examples include the following:

Polyhedra				Plane nets	3D nets	
3.4^2	3.6^2	3.8^2	3.10^2	3.12^2	3.14^2–3.20^2	3.24^2
(4.4^2)	4.6^2			4.8^2	4.12^2	4.14^2
6.4^2				(6.6^2)	6.8^2	6.10^2

The symbols for these 3-connected systems (point symbols) indicate the types of polygons meeting at each point. Examples of some of these nets are given in a later section. (An example of the net 4.10^2 does not appear to be known.)

The idea of defining uniform nets in the way just given arose in connection with 3-connected nets. It is necessary to examine whether the definition can be usefully applied to 3D nets with higher values of p. In general the number of ways of selecting two of the p links meeting at a point is $p(p-1)/2$. If $p = 3$, and only for this value, this number is equal to p, but it increases rapidly with p.

p:	3	4	5	6	$\cdots$
$p(p-1)/2$:	3	6	10	15	$\cdots$

Although a purely topological criterion for uniformity would be desirable, the requirement that all the $p(p-1)/2$ smallest circuits be n-gons is too rigorous to be useful; moreover the problem is complicated by geometrical factors. The point symbol for a uniform 3-connected net is n^3. For the 4-connected diamond net (6, 4) in which the six shortest circuits meeting at a point are 6-gons, the symbol is n^6. If the links meeting at a 4-connected point are coplanar, it is less likely that the point symbol will be of the type n^6, since the smallest circuits involving pairs of nonadjacent links are likely to be larger than those for pairs of adjacent links. For the NbO net of Fig. 9.5(4) the point symbol is $6^4 8^2$, the shortest circuits including two collinear links from a point being 8-gons. We are not aware of the existence of a 4-connected net in which there is a square coplanar arrangement of links at each point and for which the point symbol is n^6. However there are (3, 4)-connected nets in which there are points of the type n^6 at which four coplanar links meet. We therefore

Table 2.1 Values of *n* for uniform (Platonic and Catalan) 3D nets and point symbols for some Archimedean nets.*

	p					
	Type of 3D net					
	3		4		5	6
Uniform nets	n^3		n^6		n^9	n^{12}
(n, p)	7					**4**†
"Platonic"	8				**5**	
	9		**6**			
	10					
	12					
$\left(n, \begin{matrix} p \\ q \end{matrix}\right)$		3, 4		4, 6		
"Catalan"		$(n^3)(n^6)$		$(n^6)(n^{12})$		
		6				
		7				
		8		**5**		
		9				
Non-uniform nets						
$\left(\begin{matrix} m \\ n \end{matrix}, p\right)$, etc.	$\mathbf{3.20^2}$		$\mathbf{4^2.6^4}$‡		$4^6.6^3$	$\mathbf{3^6.6^6}$§
"Archimedean"	$\mathbf{4.12^2}$		$\mathbf{6^4.8^2}$			
	$\mathbf{6.8^2}$		$\mathbf{4^3.68^2}$			

* Nets with cubic symmetry in boldface type.
† Coxeter's skew polyhedron {4, 6/4}.
‡ Coxeter's skew polyhedron {6, 4/4}.
§ Coxeter's skew polyhedron {6, 6/3}.

retain our definition for 3-, (3, 4)-, and 4-connected nets. For 5- and 6-connected nets it seems reasonable to exclude the circuit(s) involving collinear bonds, therefore to require symbols n^9 rather than n^{10} for a 5-connected point and n^{12} rather than n^{15} for a 6-connected point. We describe later a 5-connected net in which the points are of the type n^9, and points of the type n^{12} exist in both a 6-connected and a (4, 6)-connected net.

Since our definition of uniformity is concerned only with the value of n, the uniform nets include those of both the Platonic and Catalan types, as shown in Table 2.1. Even when applied to nets (n, p), our condition for uniformity is not very restrictive, at least for 3-connected nets, of which we describe 30. For example, it is not necessary under the criterion that all points be topologically equivalent or that all links be equivalent, and in the next section we consider the further characterization of uniform nets by purely topological criteria. It might be expected that the selection of certain nets as *regular*, analogous to the regular configurations of the Platonic solids and plane nets (n, p), could be made according to geometrical criteria. We examine this question briefly on page 174.

In the systematic description of nets (n, p) they are grouped according to the value of p. A review in terms of n implies a more detailed look at the upper part of Table 2.1 expanded to include all systems 3^x in which all the shortest circuits are 3-gons. Since we are concerned only with uniform systems, we have to consider only systems n^3, n^6, n^9, and n^{12}. For intermediate values of the index, the complete point symbol includes other polygons as, for example,

$3^4 4^2$	$4^4 6^2$	$5^4 8^2$
Octahedron	Plane net	Plane radiating net

Horizontal shading in Table 2.2 indicates a 3D periodic net, and vertical shading indicates infinite radiating systems; these areas overlap for $6^3 6^6$ and 6^6. Systems above or to the left of the heavy line are finite.

The n^3 column is straightforward:

3^3	4^3	5^3	6^3	7^3 and so on
Tetrahedron	Cube	Pentagonal dodecahedron	Plane net	Uniform 3D nets or 2D radiating nets
	Finite			

Any (n, p) can be drawn as a plane radiating net. Rather than drawing the edges of the polygons as straight lines, it is convenient to draw these nets as

Table 2.2 Some uniform nets

	p				
n	3 n^3	3 and 4 n^3n^6	4 n^6	5 n^9	6 n^{12}
3	3^3		3^6		
4	4^3		4^6		4^{12}
5	5^3		5^6	5^9	
6	6^3 2D net	$(6^3)_a(6^6)_b$	6^6		
7	Plane radiating and uniform 3D periodic	Uniform 3D			
8					
9					
10					
12					

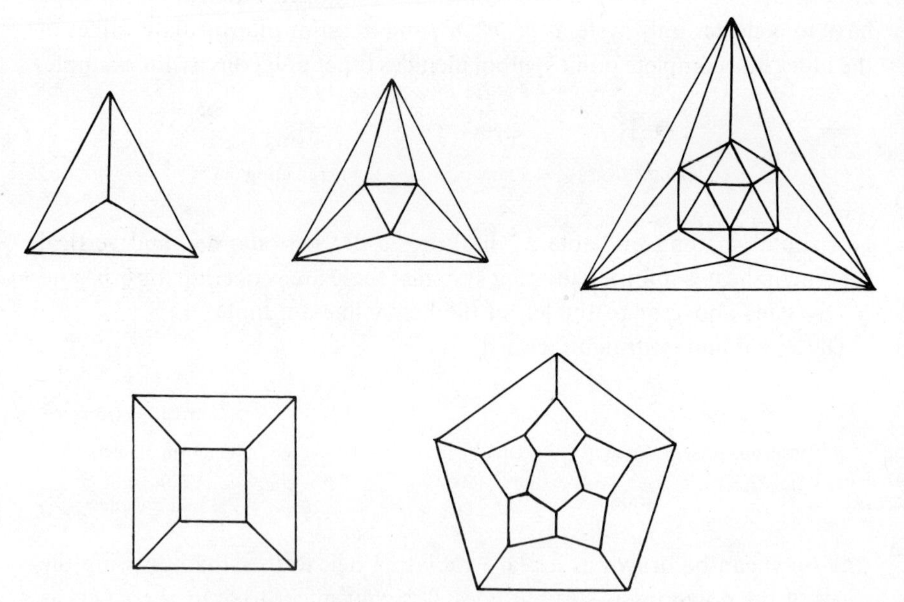

Fig. 2.1. Schlegel diagrams of the Platonic solids.

series of points on concentric circles. For (3, 3), (3, 4), (3, 5), (4, 3), and (5, 3) the nets are finite, being simply the Schlegel diagrams of the five Platonic solids (Fig. 2.1). For (3, 6), (4, 4), and (6, 3) the radiating net is identical to the periodic plane net. Since these are the only regular periodic plane nets (i.e., with all polygons n-gons and all points p-connected), all higher members of these families, namely, $\{3, p\}, p > 6, \{4, p\}, p > 4$, and $\{6, p\}, p > 3$, and their reciprocals can be realized (on the Euclidean plane) only as radiating nets. If drawn in the most symmetrical way, the highest axial symmetry is n-fold or p-fold, according to whether the center of a polygon or a point is taken as origin. The central portions of (7, 3) are drawn in these two ways in Fig. 2.2.

In the n^3n^6 column systems $(6^3)_a(6^6)_b$ include both a periodic 3D net and the radiating net of Fig. 2.3. The latter consists of an infinite series of concentric tetrahedra linked alternately by lines joining vertices and midpoints of edges. Apart from the central group of ten 3-connected points each successive tetrahedral shell contains six 3-connected and four 4-connected points, so that for the infinite system $c_3:c_4 = 3:2$. In this class there is also a 3D periodic uniform net $(6^3)(6^6)_2$ which is illustrated in Fig. 8.1. The $(n^3)_a(n^6)_b$ family continues with uniform (3, 4)-connected nets (summarized in Table 8.2) of which examples are known with $n = 7$, 8, and 9.

The n^6 column consists of

3^6 4^6	5^6	6^6
Finite	Infinite radiating net	Infinite net radiating from a line (Fig. 9.13) and uniform 3D nets (e.g., diamond)

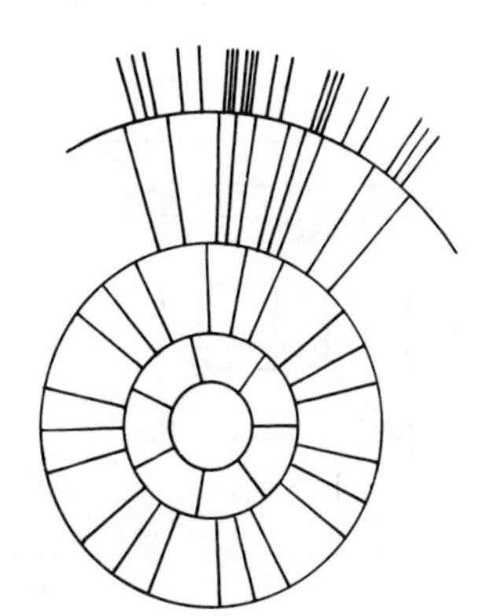

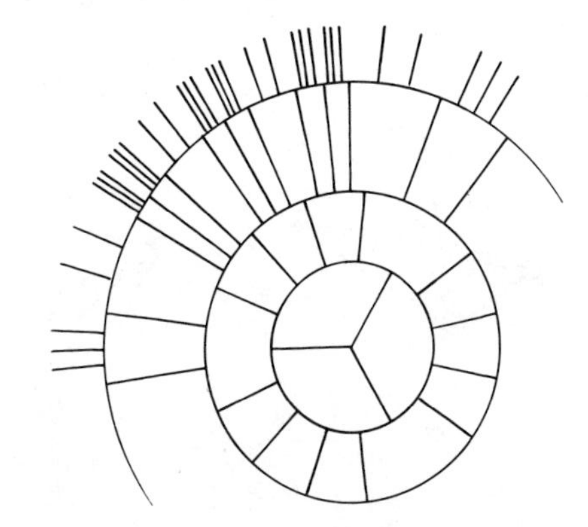

Fig. 2.2. The plane radiating net (7, 3).

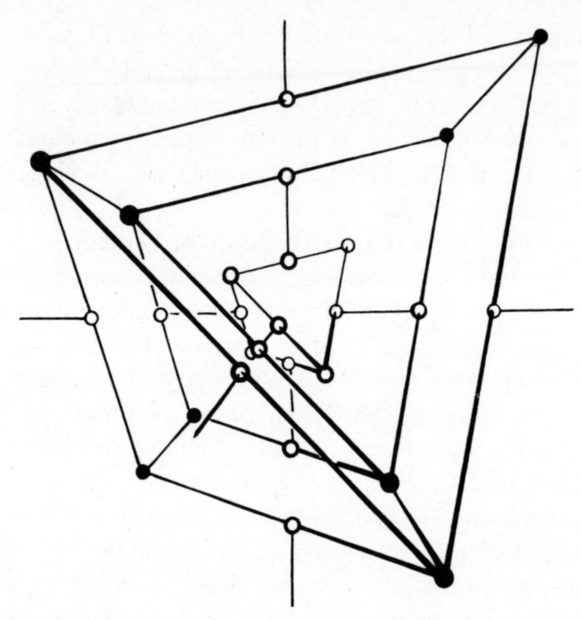

Fig. 2.3. The radiating net $\left(6, \frac{3}{4}\right)$. The sixth edge of each tetrahedron is omitted.

For $n = 3$ the system consists of a central point connected to the vertices of a tetrahedron, or the topologically equivalent case where the central point is projected through the base of the tetrahedron to form a bipyramid (Fig. 2.4*a*, *b*). For $n = 4$ each vertex of a cube is connected to one of the vertices of a circumscribing cube (Fig. 2.4*c*). In contrast to the finite 3^6 and 4^6 the system 5^6 is an infinite radiating net that starts from a central pentagonal

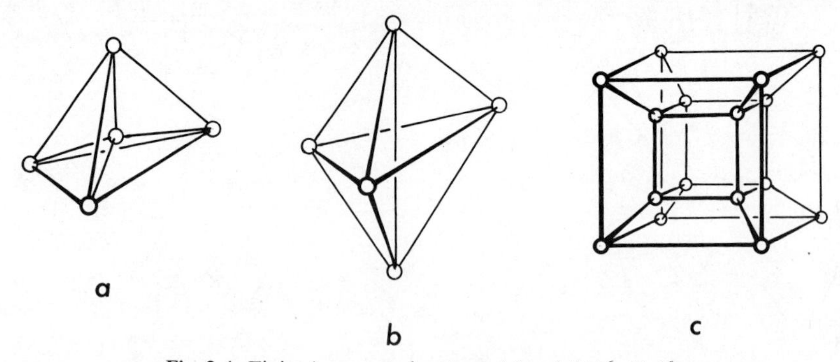

Fig. 2.4. Finite 4-connected systems. (*a*) and (*b*) 3^6; (*c*) 4^6.

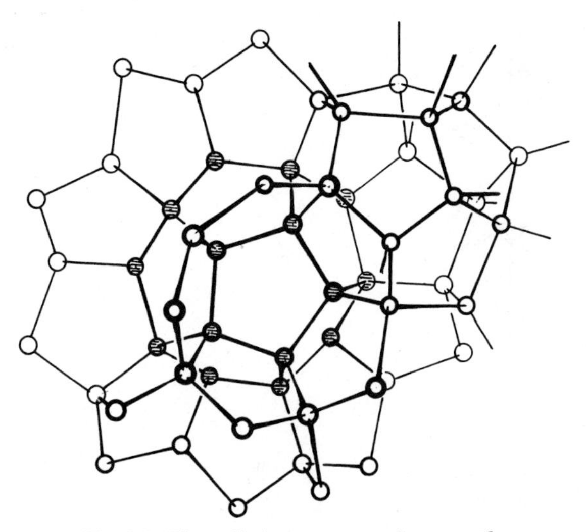

Fig. 2.5. The radiating 4-connected system 5^6.

dodecahedron (Fig. 2.5). This is surrounded by a shell of 12 dodecahedra, which is succeeded by further shells of dodecahedra.

In the remaining columns the only entries are 3D periodic uniform nets 5^9 (Fig. 16.27) and 4^{12} (primitive lattice).

The observations that the (5, 5)-4 tunnel polyhedron (p. 227) is a 3D net 5^9 and that one of the (4, 6)-6 tunnel polyhedra is 4^{12} suggested that it would be worthwhile to look for other possible nets of the types n^6, n^9, or n^{12} among the 3D polyhedra (regarded as 3D nets). It has been checked that none of the (5, 4) polyhedra (Table 16.2) is a net 5^6. Two (6, 4)-6 tunnel polyhedra will be described. The links of one are the edges of truncated octahedra in Fedorov's space-filling and the points have the symbol $6^4 4^2$. The other (Fig. 2.6) has non-equivalent points $P(6^6)$ and $Q(6^4 8^2)$; moreover, since the distance between (unconnected) points of type P is necessarily the same as that between the (connected) points Q, we should regard the net as being (4, 6)-connected rather than 4-connected. The (7, 4)-8 tunnel polyhedron contains 6-gon circuits and is therefore not a 7^6 net. None of the (4, 5) polyhedra (Table 16.2) is a net 4^9. If regarded as 3D 5-connected nets, the point symbols are (in the same order as in Table 16.2) as follows:

$$4^7 6^3 \qquad 4^5 6^5 \qquad 4^7 6^3 \qquad 4^6 6^4 \qquad 4^7 6^3 \qquad 4^7 6^3$$

If the unit of the first polyhedron (Fig. 16.22*a*) is joined to form a surface

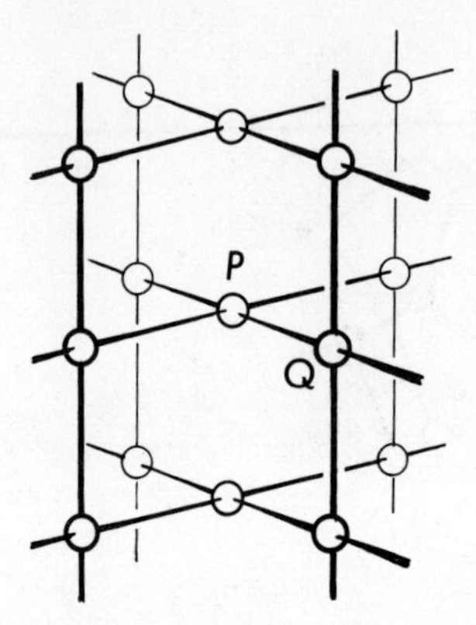

Fig. 2.6. A 4-connected net (see text); compare with Fig. 16.26B.

tessellation based on the plane 4-gon net, the point symbol is $4^8 6^2$; thus we have 5-connected nets with all the point symbols $4^5 6^5$, $4^6 6^4$, $4^7 6^3$, and $4^8 6^2$, of which the first three refer to 3D nets.

References

1. E. von Fedorov, *Z. Kristallogr.*, 1904, **38**, 321.
2. H. S. M. Coxeter, *Proc. Lond. Math. Soc.*, 1937 (2), **43**, 33.
3. A. Andreini, *Mem. Soc. Ital. Sci.*, 1907 (3), **14**, 75.

3

Topological characterization of three-dimensional nets

As soon as it was found that the symbol (n, p) does not uniquely describe a net, there being, for example, several uniform (10, 3) nets, it became desirable to characterize nets further, preferably by some purely topological properties of the nets.

In any net we can determine the number (x) of n-gons to which a point belongs and the number (y) of n-gons to which a link belongs. For a plane net $x = p$ and $y = 2$, but in 3D nets x is usually, though not necessarily, greater than p, and in some nets x has quite high values [e.g., $x = 15$ in $(10, 3) - a$]. Clearly there is a value of $x(^n x)$ for each type of polygonal circuit, but we can start by considering uniform nets and only the x value relating to the smallest (n-gon) circuits, and similarly for y; for simplicity we then omit the superscript.

All the n-gon circuits in a net (even a uniform net) are not necessarily topologically equivalent. For example, the 12-gons in (12, 3) are of two kinds (p. 33). An obvious refinement would be to distinguish between topologically nonequivalent circuits. This has not been done, and the $x(y)$ values given later in the systematic descriptions of nets refer only to the minimum circuit (n-gon). It is seen later that even the quantities n, p, $^n x$, and $^n y$ are not sufficient to characterize some uniform (n, p) nets; that is, they do not uniquely specify a single net. In such cases more detailed characterization, which has not been undertaken, would be necessary to distinguish one net of such pairs from the other.

A related point is that even in a uniform net the ${}^{n}x({}^{n}y)$ values may be identical for points (links) that are clearly topologically *non*equivalent. The nonequivalence may be demonstrated by determining ${}^{m}x({}^{m}y)$ relating to another (larger) polygon (m-gon) in the net. This point may be illustrated for the net (8, 3)-c. For this net ${}^{8}x = 6$ for all points and ${}^{8}y = 4$ for all links, but the points are clearly of two kinds, and this is also true of the links. The nonequivalence of the points is seen from the crystallographic description of the most symmetrical configuration (Fig. 5.32), but we may avoid any reference to the geometry of the net by determining the x and y values for a larger circuit such as the 18-gon circuit around a tunnel. Distinguishing the points as H and V and the links as h and v (as in Fig. 5.25) we find:

	${}^{8}x$		${}^{18}x$		${}^{8}y$		${}^{18}y$
Cell content: $2H$	6		60	$6h$	4		40
		but				but	
$6V$	6		40	$6v$	4		20
${}^{8}x_{\text{mean}}$	6	${}^{18}x_{\text{mean}}$	45	${}^{8}y_{\text{mean}}$	4	${}^{18}y_{\text{mean}}$	30

Relation between *x*, *y*, and *p*

Consider first a uniform (n, p) net in which all the points have the same ${}^{n}x$ value and all links have the same ${}^{n}y$ value; in what follows we omit the superscripts, it being understood that the values of x and y refer to the n-gon circuits. In any n-gon each edge counts as $1/y$ toward the total number of edges (links), which means that this total is $m(n/y)$ if m is the total number of n-gons in some representative portion of the net. Since each point is common to x n-gons, the total number of points is $m(n/x)$. In a p-connected net the ratio of links to points is $p/2$, therefore

$$p(n/x) = 2(n/y) \qquad \text{or} \qquad \frac{x}{y} = \frac{p}{2} \tag{3.1}$$

More generally let us suppose that in a uniform n-gon net there are points with different values of p and/or x and/or links with different values of y, and let there be, in some representative portion of the net, X_1 points associated with the value x_1, and so on, and Y_1 links with the value y_1, and so on.

At any particular point

$$x = \tfrac{1}{2}\sum y \tag{3.2}$$

since by summing the y values for all the links meeting at the point we have counted twice all the n-gons to which the point belongs. For the net as a whole, however,

$$\sum Xx = \sum Yy \tag{3.3}$$

for by summing the x values for all the points in a repeat unit of the net we have counted *twice* one-half of all the y values [equation 3.2], since every link joins two points. Weighted mean values are as follows:

$$x, \qquad x_{\text{mean}} = \frac{\sum Xx}{\sum X}$$

$$y, \qquad y_{\text{mean}} = \frac{\sum Yy}{\sum Y}$$

$$p, \qquad p_{\text{mean}} = \frac{\sum Xp}{\sum X} = \frac{2(\text{total number of links})}{(\text{total number of points})} = \frac{2\sum Y}{\sum X}$$

Therefore

$$\frac{x_{\text{mean}}}{y_{\text{mean}}} = \frac{\sum Xx}{\sum X} \cdot \frac{\sum Y}{\sum Yy} = \frac{\sum Y}{\sum X} = \frac{\text{total number of links}}{\text{total number of points}} = \frac{p_{\text{mean}}}{2}$$

Equation (3.1) therefore can be generalized to

$$\frac{x_{\text{mean}}}{y_{\text{mean}}} = \frac{p_{\text{mean}}}{2} \tag{3.4}$$

where the mean values of x, y, and p are the weighted means as defined earlier.

To illustrate this equation we give the data for the "pyrite" net, a uniform net $\left(5, \begin{matrix}4\\6\end{matrix}\right)$ in which all the shortest circuits are 5-gons and there are twice as many 4-connected as 6-connected points. In FeS_2 the former are the sulfur and the latter the iron atoms, and we use the chemical symbols for the atoms to designate the two kinds of point. The numbers of atoms (X) and of links (Y) are reduced to correspond to one formula-weight FeS_2; the numbers are 4 times larger for the (cubic) unit cell of the crystal structure. The numbers show the y values associated with the links:

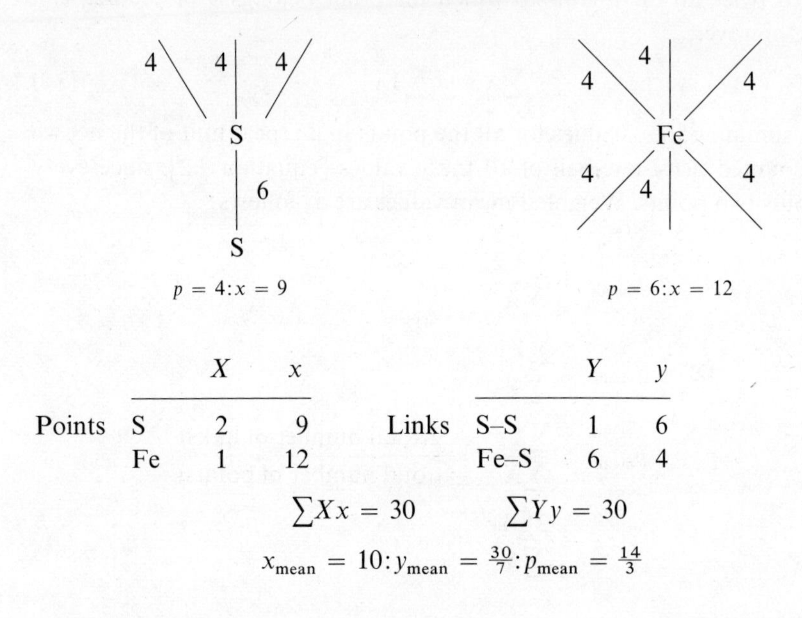

		X	x			Y	y
Points	S	2	9	Links	S–S	1	6
	Fe	1	12		Fe–S	6	4

$$\sum Xx = 30 \qquad \sum Yy = 30$$

$$x_{\text{mean}} = 10 : y_{\text{mean}} = \tfrac{30}{7} : p_{\text{mean}} = \tfrac{14}{3}$$

We record the x and y values for many uniform nets in the following pages. The values of x cover a wide range. The lowest possible value (3 for a 3-connected net) is found for a number of (8, 3) nets (Table 5.3). The highest value of x observed for points in any of the nets studied is 20, for the 4-connected points in the net $\left(8, \begin{matrix}3\\4\end{matrix}\right)$-*a*. The highest value observed for *all* points in one net is 18; in the nets of Table 3.1 the value of y is equal to n, though the significance of this is not obvious.

Table 3.1 Values of x and y for some uniform nets

p	n	Z_t	x	y	Figure
3	12	6	18	12	5.10
3	10	4	15	10	4.2(*a*)
3, 4	8	3	$13\frac{1}{3}$ (mean)	8	4.2(*b*)
4	6	2	12	6	4.2(*c*)
6	4	1	12	4	4.2(*d*)

We noted in Table 1.3 a number of 3D nets (n, p) that conform to the relation $(n - 2)(p - 2) = 8$. Table 3.2 summarizes the topological data for these nets, and all the nets are described later.

Table 3.2 Some solutions of $(n - 2)(p - 2) = 8$

n	p	${}^n y_{\text{mean}}$	${}^n x_{\text{mean}}$	3D net
3	10	$\frac{16}{5}$	16	Dual of (10, 3)-*g*
4	6	4	12	*P* lattice
5	$\frac{14}{3}$	$\frac{30}{7}$	10	Pyrite
6	4	6	12	Diamond
7	$\frac{18}{5}$	$\frac{56}{9}$	$\frac{56}{5}$	$\left(7, {3 \atop 4}\right)$
8	$\frac{10}{3}$	8	$\frac{40}{3}$	$\left(8, {3 \atop 4}\right)$ – *b* and *c*
9	$\frac{22}{7}$	$\frac{72}{11}$	$\frac{72}{7}$	$\left(9, {3 \atop 4}\right)$
10	3	10	15	(10, 3)-*a*

4

The derivation of three-dimensional nets

No equations are known for 3D nets analogous to those for plane nets relating to the proportions of polygons (circuits) of different kinds. Even in the case of plane nets the equations only give information about the relative proportions of polygons, not about the indefinitely large numbers of ways of arranging polygons of given kinds present in particular proportions. This point is illustrated later in the Chapter on 2D nets. We noted in Chapter 1 a reason for expecting to find 4-, 6-, and 10-gons in the simplest 6-, 4-, and 3-connected 3D nets; an alternative approach is as follows.

Any pattern that repeats regularly in one, two, or three dimensions consists of units that join together when repeated *in the same orientation*; that is, all units are identical and related only by translations. To form a 1D, 2D, or 3D pattern, the unit must be capable of linking to two, four, or six others, because a 1D pattern must repeat in both directions along a line, a 2D pattern along two nonparallel lines, and a 3D pattern along three (noncoplanar) lines (axes). The repeat unit may be a single point or a group of connected points, and it must have at least two, four, or six free links available for attachment to its neighbors.

Evidently the simplest unit that can form a 3D pattern is a single point forming six links, but for 4- and 3-connected 3D nets the units must contain, respectively, two and four points, as shown in Fig. 4.1. The series is obviously completed by the intermediate unit consisting of one 4- and two 3-connected

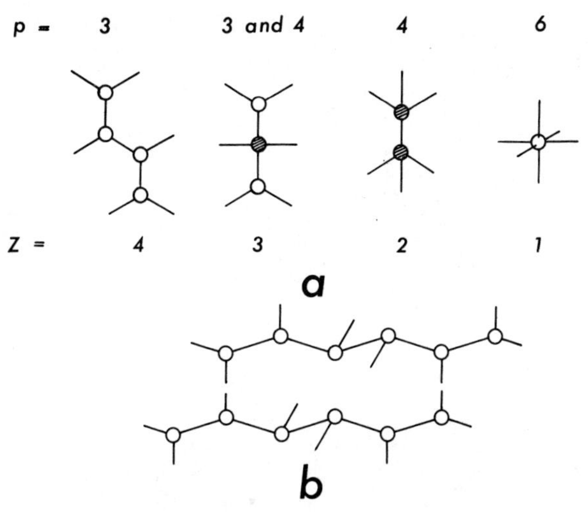

Fig. 4.1. Units forming 3D nets.

Table 4.1 Some basic 3D nets

p	n	Z_t	Z_c	y	x	Fig. 4.2
3	10	4	8	10	15	*a*
3, 4	8	3	6	8	$13\frac{1}{3}$*	*b*
4	6	2	8	6	12	*c*
6	4	1	1	4	12	*d*

* Weighted mean.

points, which also has the necessary minimum number (six) of free links. The values of Z_t, the number of points in the (topological) repeat unit, enable us to understand the simplest 3D nets. Similarly oriented units must be joined together through the free links, each one to six others. This implies that the six free links from each unit must form three pairs, one link of each pair pointing in the opposite direction to the other. Identically oriented links repeat at intervals of $(Z + 1)$ points, so that circuits of $2(Z + 1)$ points are formed. We therefore expect to find the family of basic 3D nets listed in Table 4.1, which includes the values of Z_c, the number of points in the unit cell of the most symmetrical configuration of the net, and also the values of

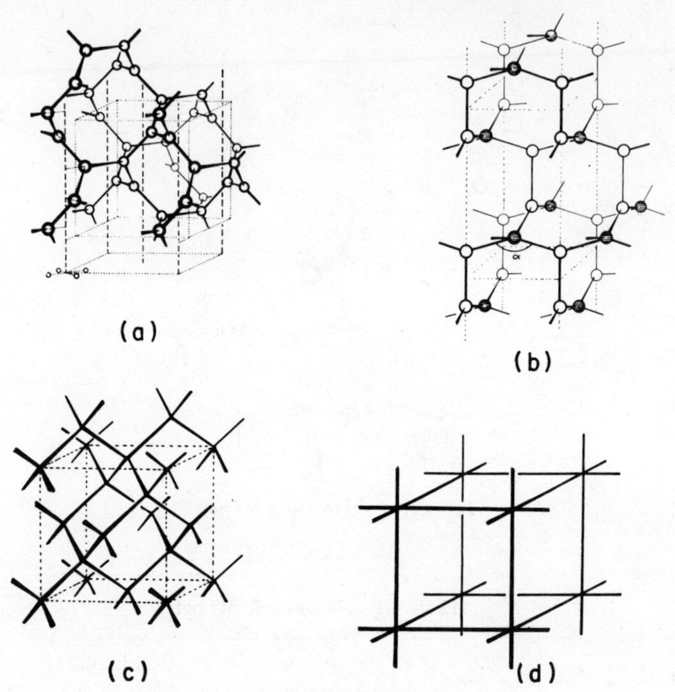

Fig. 4.2. Uniform 3D nets. (*a*) 3-Connected; (*b*) (3, 4)-connected; (*c*) 4-connected; (*d*) 6-connected.

y and *x*. Let us compare these nets, illustrated in Fig. 4.2, with the corresponding 2D nets of Figs. 1-1*a* and *b*:

p	*n*		*n*
3	10		6
		compare the	
3, 4	8	2D nets	5
4	6		4
6	4		3

3-Connected nets

The author's first attempt at the systematic derivation of 3D nets was confined to nets in which three links meet at every point. It was shown that the number of points in the repeat unit must be even and at least 4, and the first two groups of 3D 3-connected nets, with $Z = 4$ or 6, were derived in the following ways.

1. A 3D net is formed by adding 2-connected points along the links of a plane 3-connected net and joining each to a 2-connected point of an adjacent layer. Instead of adding 2-connected points, these could be formed by omitting certain links from the 3-connected plane net.
2. Alternatively we may regard some or all of the 3-, 4-, or 6-gons in plane 3-connected nets as representing helices, 3_1 or 3_2, and so on.
3. A later elaboration, giving nets with higher values of Z, was to add 2-fold screw axes perpendicular to the plane of the net, therefore parallel to the 3-, 4-, or 6-fold screw axes, and also to consider combinations of 3-, 4-, and 6-fold screw axes.

The systematic exploration of these three points is set out in Parts 1, 2, and 6 of the author's series of papers in *Acta Crystallographica*, cited in the preface, and details are not given here. The first method can be illustrated by the derivation of the nets (10, 3)-*a* and *b*. Since there must be an even number of 2-connected points in the repeat unit, 3-connected nets with $Z = 4$ can be derived only from the one plane 3-connected net having 2 points in the repeat unit (Fig. 4.3). One of the 2-connected points (A) is to be connected

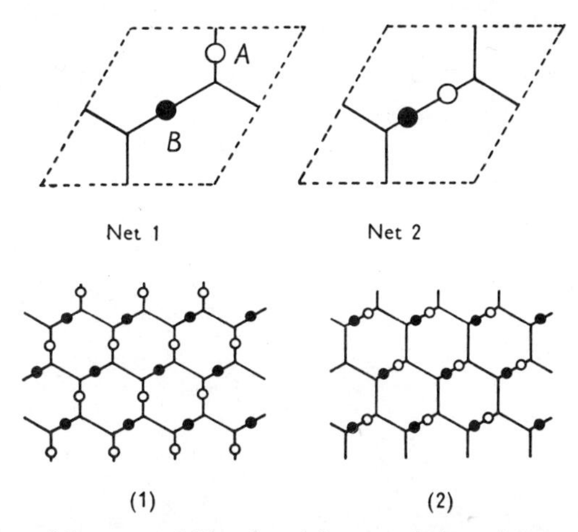

Fig. 4.3. Derivation of 3-connected Nets 1 and 2: open circle, point connected upward (U) to solid circle, point of layer above; solid circle, point connected downward (D) to open circle, point of layer below.

to B' of the layer above and the other (B) to A'' of the layer below. The nature of the nets can be more easily visualized by showing larger portions of the plane net, as indicated in the figure. The alternative derivation of (10, 3)-*a* by method (2) is illustrated in Fig. 5.2, and the projections of the nets (7, 3)-*a* and *b* in Fig. 5.42 or of (8, 3)-*d*, *g*, or *h* in Fig. 5.26 show the application of method 3.

It was found that there are only two 3-connected 3D nets with $Z = 4$; it now appears that there are 20 with $Z = 6$ (Table 4.2). We shall be interested only in certain of these nets, namely, the uniform nets and a few Archimedean nets, described in detail later. The further derivation of nets with increasing values of Z would be extremely laborious, and few of the nets are likely to be of interest. Attention therefore is focused on nets (n, p).

Table 4.2 Three-dimensional 3-connected nets

		Polygons in net									
Net	Z	3	4	5	6	7	8	9	10	11	12
1	4	—	—	—	—	—	—	—	10	—	—
2		—	—	—	—	—	—	—	10	—	12
3	6	—	—	—	—	7	—	9	10	—	—
4		—	—	—	—	—	8	—	10	—	—
5		—	—	—	—	—	8	—	—	—	12
6		—	—	—	—	—	8	—	—	—	12
7		—	—	—	6	—	—	—	10	—	12
8		—	—	—	6	—	—	—	10	—	12
9		—	—	—	6	—	—	—	10	—	—
10		—	4	—	—	—	—	—	—	—	12
11		—	—	5	—	—	—	—	—	11	—
12		—	—	—	6	—	—	—	10	—	12
13		—	4	—	—	—	8	—	—	—	—
14		3	—	—	—	—	—	—	—	—	12
15		3	—	—	—	—	—	—	—	—	12
16		—	4	—	—	—	—	—	—	—	12
17		—	4	—	—	—	—	—	—	—	12
18		—	—	5	—	—	—	—	—	11	12
19		—	—	5	—	—	—	—	—	11	12
20		—	—	—	—	—	8	—	10	—	12
21		—	—	—	—	—	—	—	—	—	12
22		—	—	—	—	—	—	—	10	—	12

Two other sources of 3-connected 3D nets can be noted here. The duals of triangulated surface tessellations (3, p) provide numerous (n, 3) nets, some with rather high values of Z; those which are uniform nets (and on this point see p. 261) are described later. The last method derives 3-connected 3D nets from other 3D nets. Replacement of some or all of the points in a uniform 3-connected net by triangles gives Archimedean nets; for example, a uniform (10, 3) net gives a net with the point symbol 3.20^2 if all points are replaced by triangles. On the other hand, replacement of some or all points in a 4-connected net by either a pair of 3-connected points or by a 4-gon gives either a (3, 4)- or a 3-connected net. Examples of such relations between nets are given in Table 4.3. Different nets may result from such replacements, depending on the orientation of the unit replacing the 4-connected point. For example, the diamond net gives the net (10, 3)-*a* if cubic symmetry is retained or the net (10, 3)-*b* if all the pairs of 3-connected points are similarly oriented.

Table 4.3 Relations between some 3D nets

Replacement of × or + by >—<	Replacement of × or + by >▭<
Diamond → (10, 3)-*a*; Diamond → (10, 3)-*b*	Diamond → 4.14^2
NbO → (9, 3)-*a*	NbO → 4.12^2
$\left(8, \frac{3}{4}\right)$-*a* → (10, 3)-*g*	
$\left(8, \frac{3}{4}\right)$-*d* → (9, 3)-*b*	

Because the (n, 3) nets have been derived by a variety of methods rather than by the continued application of one method, it cannot be claimed that our enumerations of uniform nets are complete up to the highest Z value given for a particular family. The information contained in the summarizing tables given later, however, could provide the starting point for further studies.

(3, 4)-Connected nets

No attempt has been made to derive (3, 4)-connected nets systematically. A number of uniform (3, 4)-connected nets are described later, including those intermediate between the diamond net and the (10, 3) nets; they are of special interest as examples of 3D nets in which all the shortest circuits are 7-, 8-, or 9-gons, respectively.

4-Connected nets

The systematic study of some of the simpler 4-connected nets is described, together with descriptions of 4-connected nets of greater complexity, some of which are the bases of crystal structures.

3D nets, $p > 4$

A short section on more highly connected nets is included later; no systematic studies are described.

5

Uniform 3-connected nets

Two uniform (8, 3), one (9, 3), three (10, 3), and one (12, 3) net were derived in Part 1, the last being incorrectly symbolized by symbol $12^2.14$. In that paper a less restrictive definition of uniformity was suggested, which was later discarded in favor of the one adopted in the present account. Further examples of uniform nets were derived in Part 6, where two uniform (7, 3) nets were described. In Part 7 additional uniform (n, 3) nets emerged as the duals of 3D polyhedra, and some were illustrated. The relatively simple uniform net (10, 3)-c of Table 5.1 was not found in the earlier study of 3-connected nets having $Z_t = 6$; it was shown later to be the basis of the structure of crystalline B_2O_3. The known uniform 3-connected nets now number 30:

Uniform nets	Number
(12, 3)	1
(10, 3)	7
(9, 3)	3
(8, 3)	15
(7, 3)	4

They are now described in that order.

The net (12, 3)

If a 6_1 helix is set up at each corner of a hexagonal unit cell and a second helix starts $d/2$ out of phase with the first (d being the repeat distance along the helical axis), the helices can be joined together as shown in projection in

Fig. 5.1*a** and as a stereo-pair in Fig. 5.10. All the points form one continuous net in which the smallest circuits are 12-gons. The most symmetrical (hexagonal) configuration of this net is described as follows:

Hexagonal Space group $P6_222$ (No. 180) $6(g)$ $(x00)$

Since there is a variable x parameter (in addition to c/a), there is no unique configuration of this net, and the condition for three (and only three) equidistant nearest neighbors is that the values of c/a and x correspond to points on the plot of $c/a = 9(1 - 4x + 3x^2)$ between the intersections with the lines $x = \frac{1}{4}$ and $c/a = 1 - 2x$.

The topology of the net is as follows. All points have $x = 18$, the highest value for any 3-connected net. The links are of two kinds, those in the helices (h) having $y = 11$ and the links (j) joining the helices (which are one-half as numerous as the former) having $y = 14$, giving $y_{\text{mean}} = 12$. The 12-gon circuits are of two kinds, shown in projection in Figs. 5.1*b* and *c*. For brevity we refer to these as type *b* or type *c* 12-gons. The former resemble an open knot, and since they consist of alternate h and j links it follows that the 12-gon circuit including two h links from any point cannot be of type *b* (refer to Fig. 5.1). The 18 12-gons to which any point belongs are made up of six of type *b* and twelve of type *c*; the y values are as follows:

h link belongs to 3 12-gons of type *b* and 8 of type *c*: total $y = 11$

j link belongs to 6 12-gons of type *b* and 8 of type *c*: total $y = 14$

Thus although this (12, 3) net has the highest observed value of x for a 3-connected net and the largest value of n, it is of lower topological (and crystallographic) symmetry than the cubic net (10, 3)-*a*. All the helices in (12, 3) are clockwise (or all anticlockwise), and the net is therefore enantiomorphic.

We have described this net in some detail because of its special features and also because it was incorrectly described in Part 1 as a net with point symbol $12^2.14$. It cannot be constructed with all bond angles equal to 120°.

This (12, 3) net is closely related to a 6.10^2 net described in Chapter 6. If the two intertwining 6_1 helices are connected as in Fig. 5.1*d*, closed (puckered) hexagons are formed. Whereas Fig. 5.1*a* represents a single (12, 3) net, the projection of Fig. 5.1*d* represents two identical interpenetrating 6.10^2 nets that are not connected to each other.

* In some figures representing projections of nets involving screw axes, the heights of points should refer to distances *below* the plane of the paper to conform to the conventional descriptions of screw axes; alternatively, for 3_1 read 3_2, for 4_1 read 4_3, and so on.

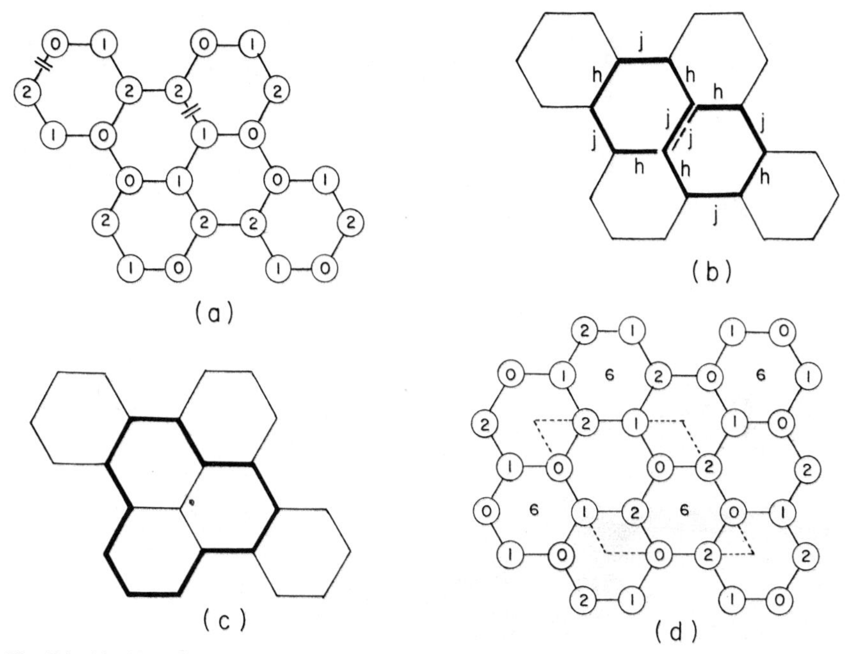

Fig. 5.1. (*a*)–(*c*) Projections of net (12, 3) on (001). Numerals in (*a*) show the heights of points as multiples of $c/3$, where c is the height of the hexagonal cell. (*d*) Projection of two interpenetrating 6.10^2 nets (p. 82).

Uniform (10, 3) nets

The seven known uniform (10, 3) nets are listed in Table 5.1.

The net (10, 3)-a

This net occupies a unique position among 3D 3-connected nets. It has the highest x and y values of any (10, 3) net—15 and 10, respectively, all points are equivalent, all links are equivalent, and all 10-gons are congruent. In its most symmetrical configuration it has cubic symmetry, all links of equal length, and all interbond angles 120°:

Cubic Space group $I4_132$ (No. 214) $Z = 8$ in $8(a)$ $(\frac{1}{8}\frac{1}{8}\frac{1}{8})$

The unit cell is illustrated in Fig. 5.2*a*, and projections along the 4_1 and 3-fold axes are in Figs. 5.2*b* and *c*, respectively. Figure 5.11 shows a stereo-pair of a wire model.

Table 5.1 Uniform (10, 3) nets*†

Net	Z_t	x		y	y_{mean}	Examples
a‡	4	15		10	$\frac{30}{3}$	H_2O_2, $Hg_3S_2Cl_3$, $SrSi_2$, $CsBe_2F_5$, Sn_2F_3Cl
b‡	4	10		8, 6, 6	$\frac{20}{3}$	$ThSi_2$, P_2O_5, $La_2Be_2O_5$, $(Zn_2Cl_5)(H_5O_2)$, $(CH_3)_2S(NH)_2$
c‡	6	5		4, 3, 3	$\frac{10}{3}$	B_2O_3
			x_{mean}			
d	8	10	—	8, 6, 6	$\frac{20}{3}$	α- and β-resorcinol
e	12	9, 11	10	8, 6, 6	$\frac{20}{3}$	
f	16	10	—	8, 6, 6	$\frac{20}{3}$	
g	20	12	—	8	8	

* The net e has equal numbers of two kinds of nonequivalent point, and the nets b–f inclusive have two kinds of nonequivalent link, one-third of one kind and two-thirds of the other. Note that the number of links per cell is equal to $3Z_t/2$ and that the numbers given in the y column refer to the three links meeting at each point.

† Symbols as follows: Z_t = number of points in smallest cell (topological),
x = number of 10-gon circuits to which each point belongs,
y = number of 10-gon circuits to which each link belongs.

‡ Can be constructed with all bonds equal in length and all bond angles 120°.

Like (12, 3), the net (10, 3)-a is enantiomorphic, and the possibility of forming a novel type of racemate from two interpenetrating nets, one D and the other L) is mentioned under "Interpenetrating Nets" (Chapter 11). In the chapter "Relation of 3D Nets to Sphere Packings" we note that the points of this net are the centers of the spheres in one of Heesch and Laves' (H and L) open sphere packings, and that four (10, 3)-a nets can interpenetrate to form cubic closest packing.

This net can be derived from the diamond net by replacing each 4-connected point by a pair of 3-connected points in such a way as to maintain cubic symmetry. We refer in Chapter 13 to a second special configuration of (10, 3)-a with mutually perpendicular links. Replacement of the points of this net by equilateral triangles gives the net 3.20^2 described on page 75. The edge-sharing octahedral AX_3 structure based on this net is illustrated in Fig. 7.3.

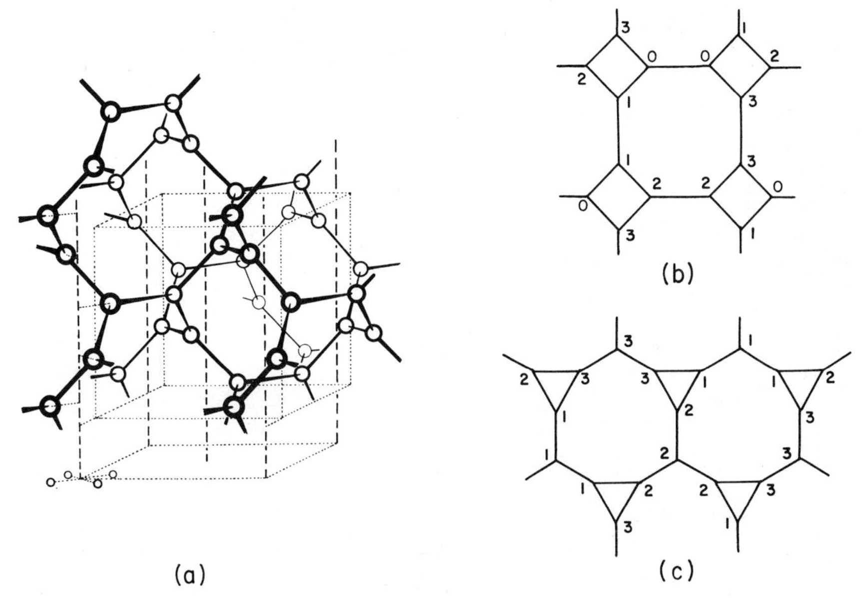

Fig. 5.2. The uniform net (10, 3)-*a*. (*a*) Unit cell; (*b*) projection along 4-fold axis; (*c*) projection along 3-fold axis. Numerals indicate heights of points as multiples of $\frac{1}{4}$ or $\frac{1}{3}$ of the repeat distance along the 4_1 or 3_1 axis.

The net (10, 3)-b

This net may be derived by replacing the points in the diamond net by pairs of 3-connected points if the links between the latter are all parallel to one 4-fold axis. This gives the most symmetrical configuration of this net, which is described as follows:

Tetragonal Space group $I4_1/amd$ (No. 141) $c/a = 2\sqrt{3}$

$Z = 8$ in. $8(e)$ $(00z)$ $z = \frac{1}{12}$

In this configuration of the net all bonds are equal in length and all interbond angles are 120°. A unit cell is illustrated in Fig. 5.3 and a stereo-pair in Fig. 5.12. All the points are equivalent ($x = 10$), but the links are of two kinds, those meeting at any point having $y = 8$ (one) and $y = 6$ (two), giving $y_{\mathrm{mean}} = 6\frac{2}{3}$. The edge-sharing octahedral structure based on this net is illustrated in Fig. 7.4, and other configurations of this net are described in Chapters 11, 12, and 13.

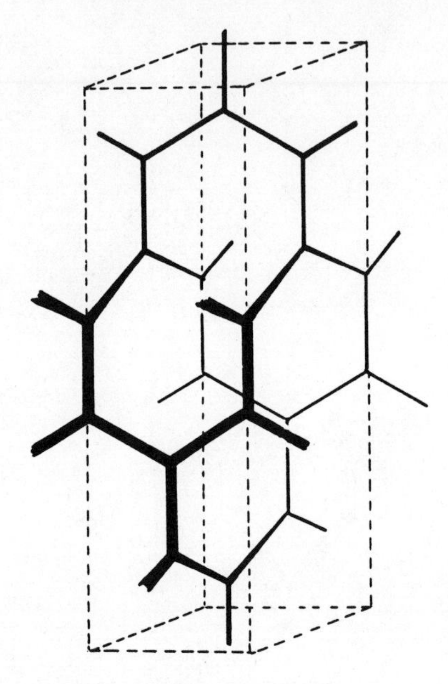

Fig. 5.3. Unit cell of tetragonal configuration of the net (10, 3)-*b*.

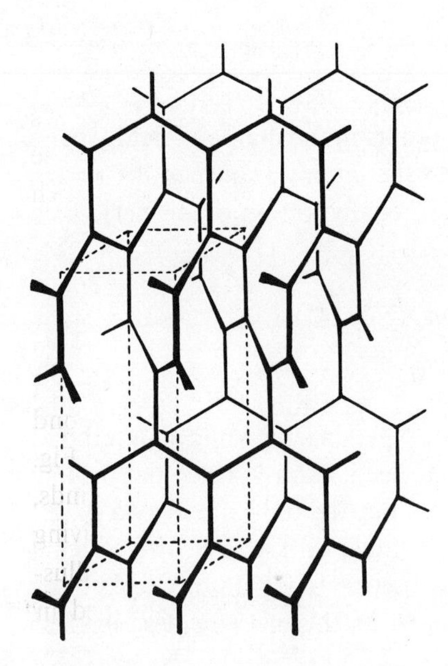

Fig. 5.4. The net (10, 3)-*c*. (The unit cell is not that of the International Tables but has its origin at $(\frac{1}{3}, \frac{1}{6}, \frac{1}{6})$.)

The net (10, 3)-c

This net is closely related to the preceding one, but the zigzag chains from which it may be built are related by successive rotations through 60° (or 120°) rather than 90° as in the net (10, 3)-*b*. The most symmetrical configuration is fully described as follows:

Hexagonal Space group $P3_112$ (No. 151) $c/a = (3\sqrt{3})/2$

$Z = 6$ in. $6(c)$; $x = \frac{1}{3}$; $y = \frac{1}{6}$; $z = \frac{1}{9}$

In this configuration of the net (Fig. 5.4) all links are equal in length and all interbond angles are 120°. The net is illustrated stereoscopically in Fig. 5.13. As in (10, 3)-*b* the links are of two kinds. Those parallel to the *c* axis have $y_c = 4$, whereas those in the zigzag chains have $y_a = 3$. Since the former constitute one-third of the total, the value of y_{mean} is $\frac{10}{3}$. This net is evidently the third member of a family of uniform (10, 3) nets that have *x* equal to 15, 10, and 5, respectively. For the relation of this net to (10, 3)-*b* and to the planar 6-gon net see Appendix to this chapter.

The nets (10, 3)-d and (10, 3)-f

These two nets are closely related to each other and to the cubic net (10, 3)-*a*. All three project as the plane 4.8^2 net; they differ in the arrangement and type of 4-fold screw axes. As a result of the way in which the helices are interconnected in these nets (Fig. 5.5); successive links in both the 4_1 and 4_3

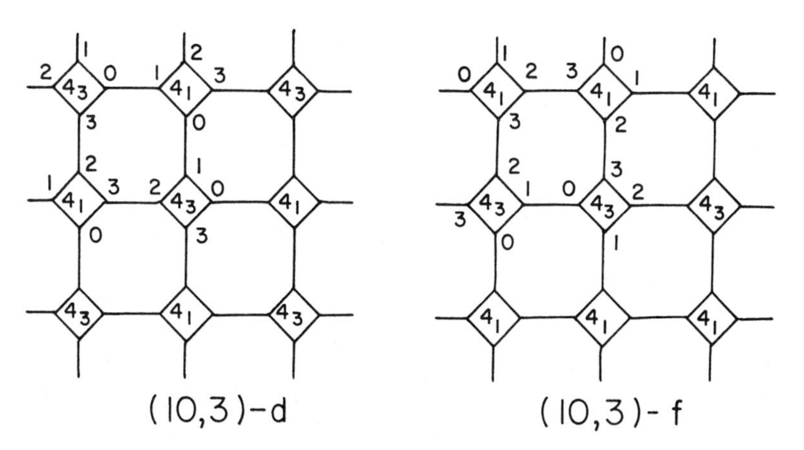

Fig. 5.5. Projections of the nets (10, 3)-*d* and (10, 3)-*f*.

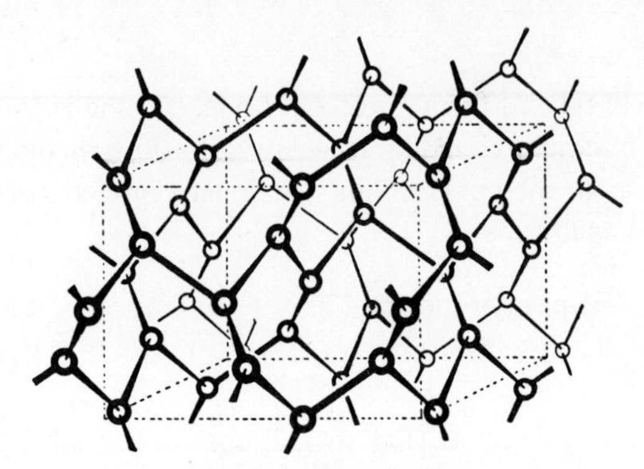

Fig. 5.6. The net (10, 3)-*d*.

helices are topologically nonequivalent and have y values of 6 and 8. Since the cross-links have $y = 6$, all points have $x = 10$. Neither of these nets has a symmetrical configuration with equal links and 120° bond angles.

The most symmetrical configuration of (10, 3)-*d* is described as follows:

Orthorhombic $Z = 8$ Space group *Pnna* (No. 52)

8 points in the general position (xyz)

Two configurations of this net represent the basic topologies of the structures of the two polymorphs of resorcinol (see Tables 5.1 and 13.2), the space group of both of which is $Pn2_1a$ (No. 33). This net is illustrated in Fig. 5.6. The net (10, 3)-*f* is illustrated stereoscopically in Fig. 5.15.

The net (10, 3)-e

This net arises by erecting 6_1 axes at the points of the plane (3, 6) net and 2_1 axes at the midpoints of the links. One unit cell of the projection is shown in Fig. 5.7. This net cannot be constructed with equal bonds and bond angles of 120°. The model illustrated as a stereo-pair in Fig. 5.14 is built with bonds in the 2_1 axes longer than those in the 6_1 axes. This net is enantiomorphic; a net in which 6_1 and 6_5 axes alternate is not possible because the polygons in the original plane net have an odd number of sides.

The net (10, 3)-g

This net was derived as the dual of the 3D polyhedron (3, 10), of which a repeat unit is shown in Fig. 16.19*b*. On this surface (of varying curvature) the edges of the triangles are of equal length. The corresponding 3D polyhedron with plane triangular faces may be visualized in the following way.

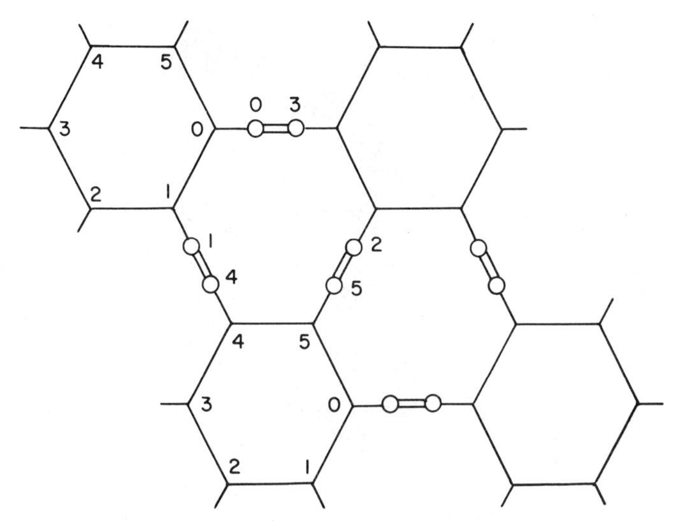

Fig. 5.7. Projection of the net (10, 3)-*e*.

The 12 points of the type ($\frac{1}{2}0\frac{1}{4}$), ($\frac{1}{2}0\frac{3}{4}$), ($0\frac{1}{4}\frac{1}{2}$), ($0\frac{3}{4}\frac{1}{2}$), ($\frac{1}{4}\frac{1}{2}0$), ($\frac{3}{4}\frac{1}{2}0$), on the faces of a cube are the vertices of a "cubic icosahedron" having eight equilateral and six pairs of isosceles triangular faces. If a centrosymmetric set of six of the latter faces are omitted and additional isosceles triangles are added to form tunnels connecting each icosahedron to six others, we have a plane-faced version of the 3D polyhedron (Fig. 5.8). The polyhedron fills one-half of space, the space not occupied by the polyhedron ("complementary polyhedron") being identical with the polyhedron itself. Alternatively, the polyhedron derives from a space-filling arrangement of icosahedra and isosceles tetrahedra (in the ratio 1:3), by removing one-half of each kind of polyhedron and removing also the faces shared between the icosahedra and tetrahedra so that there is a continuous connected space within the 3D polyhedron. This polyhedron has a unit cell with edge length twice that of Fig. 5.8 and the space group is $Ia3$. The model in Fig. 16.21 was constructed from equilateral triangles of edge length l and isosceles triangles of edge length l (two) and $2l/\sqrt{6}$. The cell contains 48 vertices, that is, 8 times the value given on page 223 for the topological repeat unit. The dual, a (10, 3) net, is formed by joining the centers of the 10 triangles meeting at each vertex, thus forming 10-gons around the vertices of the original polyhedron. The A points are the centers of the equilateral triangles (with coordinates of the type $\frac{1}{4}\frac{1}{4}\frac{1}{4}$ in the subcell). The B points are the centers of the isosceles triangles, that is, the points of intersection of the perpendicular bisectors of the edges. The unit cell contains

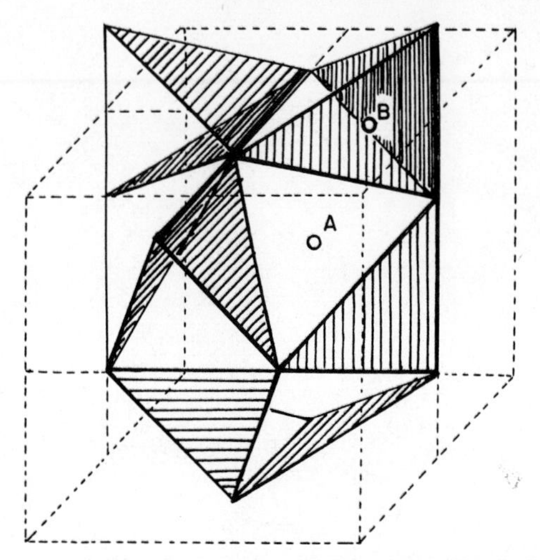

Fig. 5.8. Relation of the net (10, 3)-*g* to the 3D polyhedron (3, 10). The shaded isosceles triangles are those that form the sides of the tunnels in the 3D polyhedron.

160 points [i.e., (10/3)(48)], 64*A* and 96*B*:

Space group *Ia*3 (No. 206)

A points in 16(*c*)	$x = \frac{1}{8}$
48(*e*)	$x = \frac{1}{8}, y = \frac{1}{8}, z = \frac{3}{8}$
B points in 48(*e*)	$x = \frac{1}{10}, y = 0, z = \frac{1}{5}$
48(*e*)	$x = \frac{7}{20}, y = \frac{1}{20}, z = \frac{1}{4}$

The links in this net are not equal in length, being respectively 0.074a (A–B) and 0.05a (B–B), where a is the length of the cell edge.

For convenience in building the model (Fig. 5.16) and also to illustrate the topological relation of this net to the (3, 4)-connected net of Fig. 8.9, we have used slightly different coordinates for the B points, which correspond to replacing the 4-connected points in Fig. 8.9 by pairs of 3-connected points in the manner shown in Fig. 5.9.

There are two kinds of topologically nonequivalent point, both with ${}^{10}x = 12$, and two kinds of nonequivalent link, both with ${}^{10}y = 8$. Presumably differences between the A and B points and between the A–B and B–B links would show up if the x and y values relative to some larger polygons (e.g., ${}^{12}x$ and ${}^{12}y$) were determined. This point has not been checked.

Figures 5.10 through 5.16 are stereo-pairs of uniform (12, 3) and (10, 3) nets.

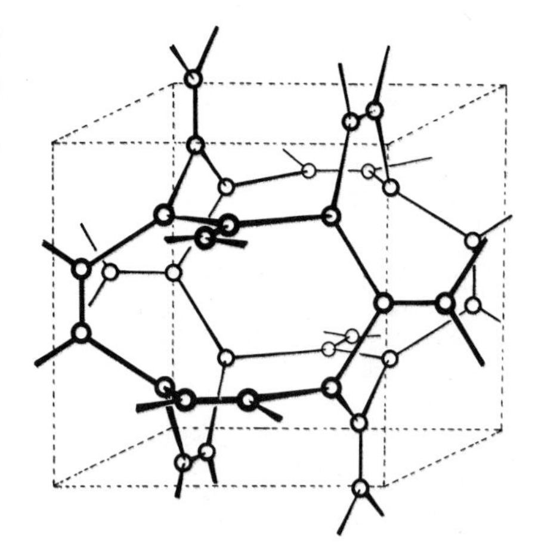

Fig. 5.9. Derivation of the net (10, 3)-*g* by replacing the 4-connected points in $\left(8, \begin{matrix}3\\4\end{matrix}\right)$-*a* by pairs of 3-connected points (cf. Fig. 8.9). Owing to the orientations of the links joining the pairs of 3-connected points, the unit cell of the (3, 4)-connected is a subcell of the 3-connected net. Doubling of each cell dimension raises the cell content from 20 to 160 points.

Fig. 5.10. The uniform net (12, 3).

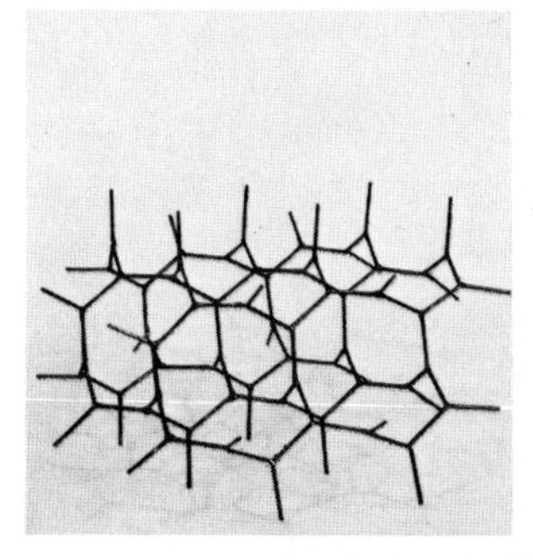
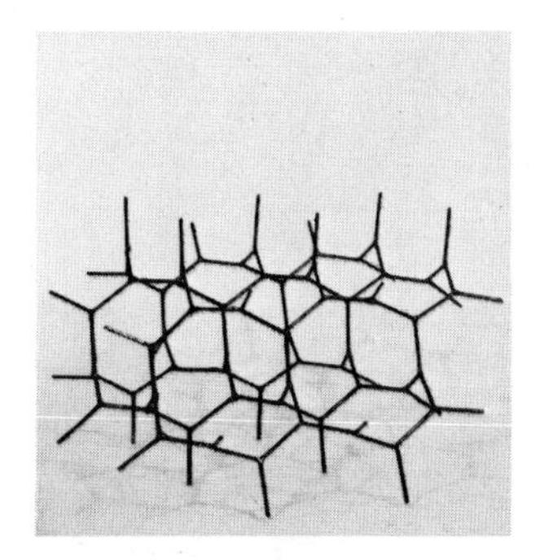

Fig. 5.11. The uniform net (10, 3)-*a*.

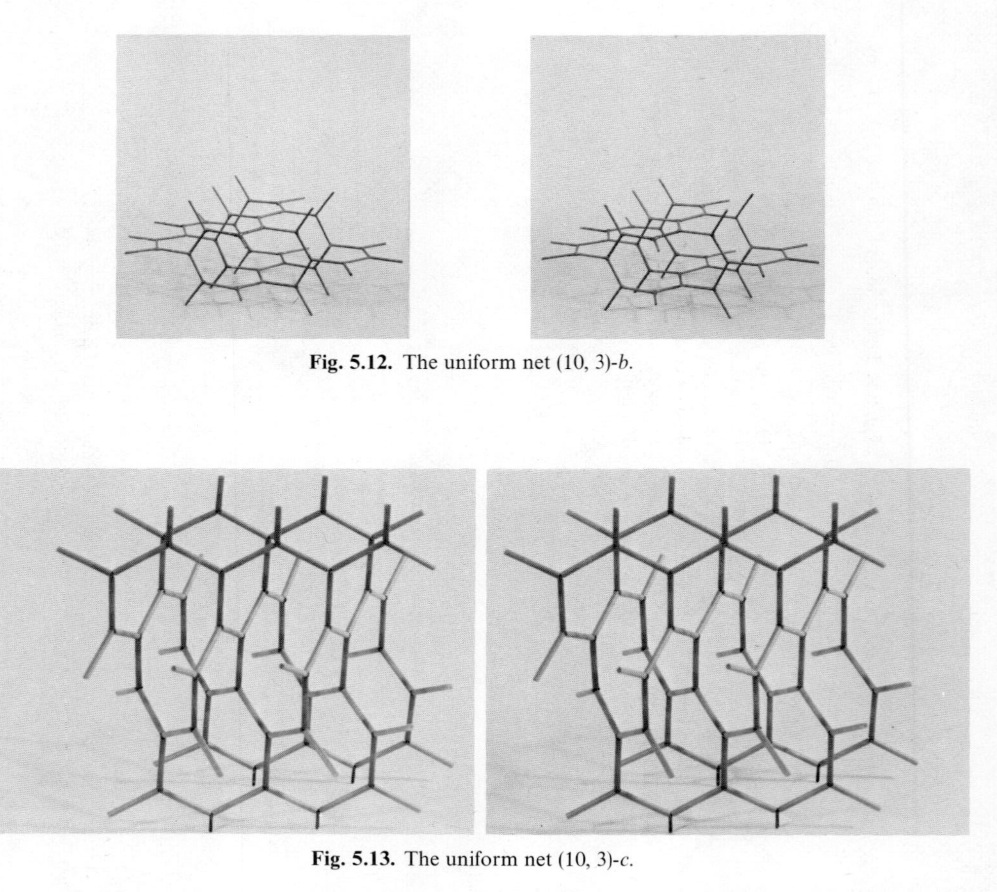

Fig. 5.12. The uniform net (10, 3)-*b*.

Fig. 5.13. The uniform net (10, 3)-*c*.

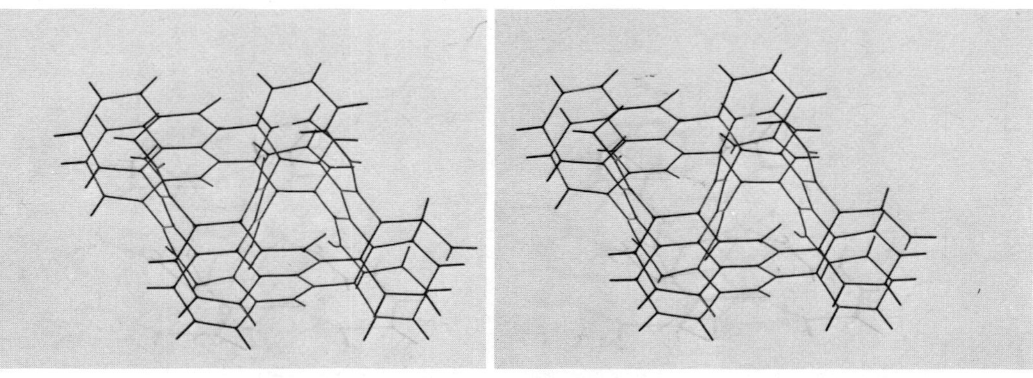

Fig. 5.14. The uniform net (10, 3)-*e*.

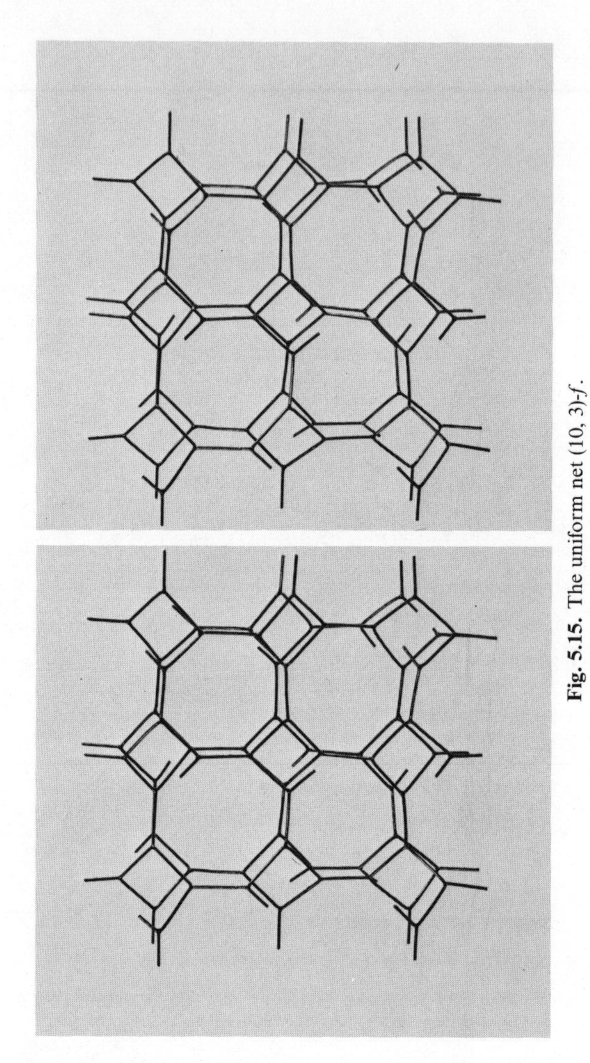

Fig. 5.15. The uniform net (10, 3)-*f*.

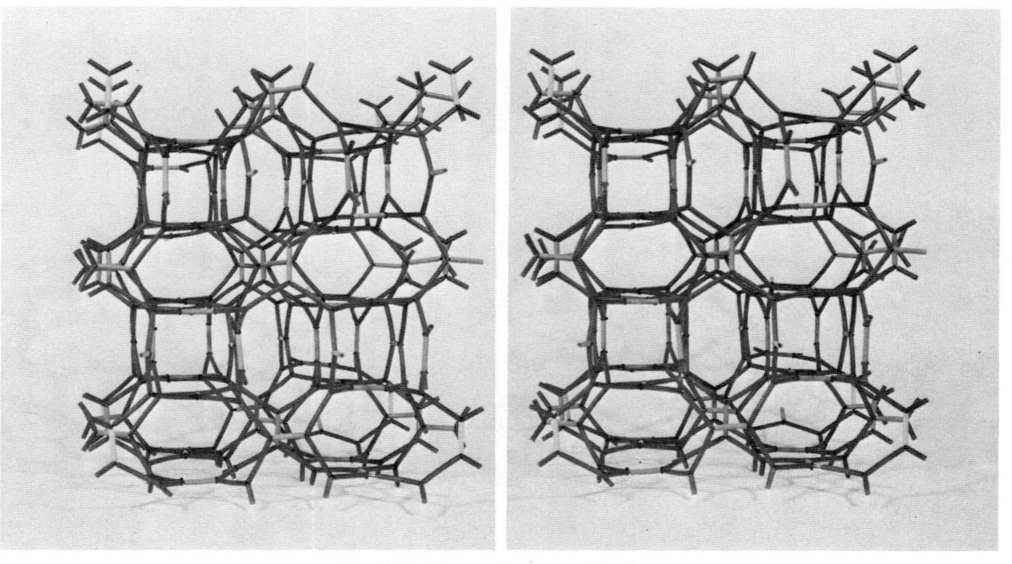

Fig. 5.16. The uniform net (10, 3)-*g*.

Uniform (9, 3) nets

The net (9, 3)-a

This net was first derived by joining together layers of the plane 6-gon net as shown in Fig. 5.17*a*. It is alternatively derived from the plane (3, 4)-connected 5-gon net by erecting 4_{12} axes (see Part 6) at the 4-connected points, as in Fig. 5.17*b*; this is the projection of this net along the direction $[\bar{1}101]$. This net can be constructed with equal bonds and all interbond angles 120° when it is completely described as follows:

$$\text{Rhombohedral} \qquad \text{Space group } R\bar{3}m \text{ (No. 166)} \qquad Z = 12$$

$$\text{Hexagonal} \qquad Z = 36 \qquad c/a = \frac{\sqrt{6(4+\sqrt{3})}}{1+2\sqrt{3}} = 1.314$$

In the hexagonal cell the points are

$$18(f)\text{:} \quad xyz; \quad x = \frac{\sqrt{3}}{1+2\sqrt{3}} = 0.388; \quad y = 0; \quad z = 0$$

$$18(h)\text{;} \quad x0z; \quad x = \frac{1+\sqrt{3}}{4(1+2\sqrt{3})} = 0.153; \quad z = \tfrac{3}{4}$$

It is illustrated stereoscopically in Fig. 5.21, which shows rather more than one unit cell. There are equal numbers of two kinds of nonequivalent point, both having $x = 6$, and three kinds of nonequivalent link, all having $y = 4$.

This net is related to the 4-connected NbO (6-gon) net, from which it can be derived by replacing the 4-connected points by pairs of 3-connected points.

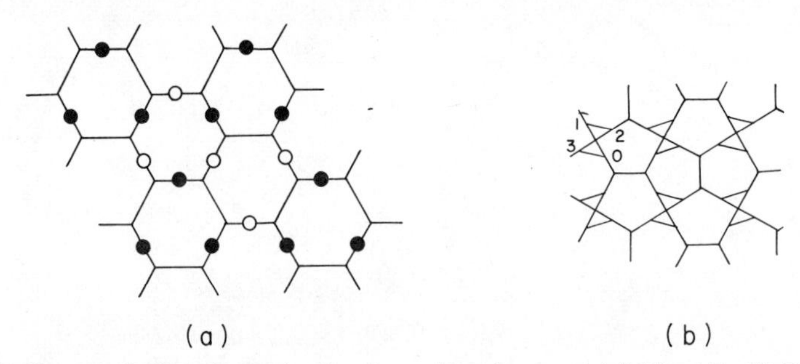

Fig. 5.17. The net (9, 3)-*a*. (*a*) Derivation from plane 6-gon net (cf. Fig. 4.3). (*b*) Derivation from (3, 4)-connected 5-gon net.

The links between the pairs replacing niobium atoms are approximately parallel to one 3-fold axis of the cubic NbO net, and the links between the pairs replacing oxygen atoms are perpendicular to the 3-fold axis. Topological data for uniform (9, 3) nets are given in Table 5.2.

Table 5.2 Uniform (9, 3) nets*

Net	Z_c	x	x_{mean}	y	y_{mean}
(9, 3)-*a*	12	6 (six)	6	4 (three)	4
		6 (six)		4 (three)	
				4 (twelve)	
(9, 3)-*b*	24	6 (eight)	6	4 (four)	4
		6 (sixteen)		4 (sixteen)	
				4 (sixteen)	
(9, 3)-*c*	24	7 (sixteen)	$7\frac{1}{2}$	6 (twelve)	5
		8 (four)		5 (twelve)	
		9 (four)		4 (twelve)	

* The x and y values are given for the various sets of nonequivalent points (total Z) and nonequivalent links (total $3Z/2$). The data for the nets (9, 3)-*a* and *b* refer to the rhombohedral and tetragonal cells, respectively. For (9, 3)-*b* the value of Z is also that of Z_t, this net being the dual of a (3, 9)-6*t* polyhedron for which $Z_t = 8$.

The net (9, 3)-b

This net is the dual of the 6-tunnel (3, 9) 3D polyhedron illustrated in Fig. 16.19*a*. Alternatively it may be derived from the uniform net $\left(8, \begin{matrix}3\\4\end{matrix}\right)$-*d* by replacing the 4-connected points by pairs of 3-connected points, successive units along the c axis being rotated through 90°—compare the relation between (10, 3)-*g* and $\left(8, \begin{matrix}3\\4\end{matrix}\right)$-*a*. The relation to the net $\left(8, \begin{matrix}3\\4\end{matrix}\right)$-*d* can be seen from the projection of Fig. 5.18. Since the columns projecting as squares may be rotated through 90° relative to one another, there is an indefinitely large number of (9, 3) nets forming a family like the (10, 3) nets that arise from different combinations of parallel 4_1 and 4_3 axes.

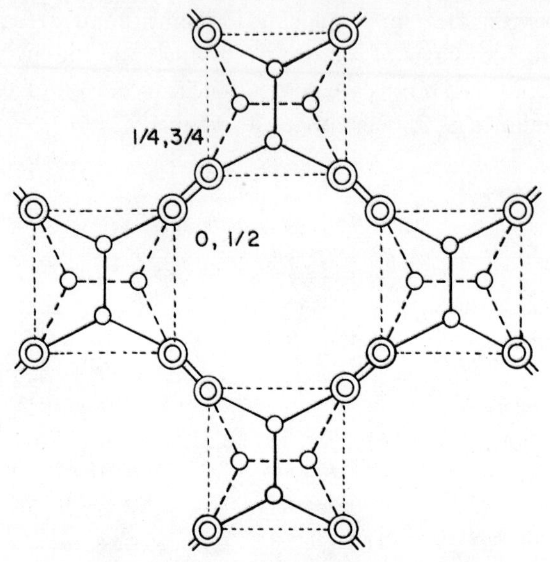

Fig. 5.18. Projection of the net (9, 3)-*b*.

Like the net (9, 3)-*a*, the net (9, 3)-*b* can be constructed with all bonds equal in length and all interbond angles 120°. This highly symmetrical configuration of the net is illustrated in Fig. 5.19 as a projection along the tetragonal *c* axis and as a stereo-pair in Fig. 5.22. It is fully described as follows:

Tetragonal Space group $P4_2/nmc$ (No. 137; origin at $\bar{4}m2$) $Z = 24$

$$c/a = \tfrac{1}{2}(8\sqrt{3} - 11)^{1/2} = 0.845$$

$$8(g): \quad 0xz; \quad x = \tfrac{3}{8}; \quad z = \tfrac{1}{8}$$

$$16(h): \quad xyz; \quad x = \frac{\sqrt{3}}{8} = 0.2165; \quad y = \tfrac{1}{4}; \quad z = \tfrac{1}{8}$$

Although there are two sets of nonequivalent points, all have $x = 6$; all three sets of nonequivalent links have $y = 4$. These x and y values are the same as for the preceding net.

The net (9, 3)-c

There is no configuration of this net with equal coplanar bonds. The net appears in projection in Fig. 5.20*a* and as a stereo-pair in Fig. 5.23. Topologically this net is much more complex than the preceding ones, there being three

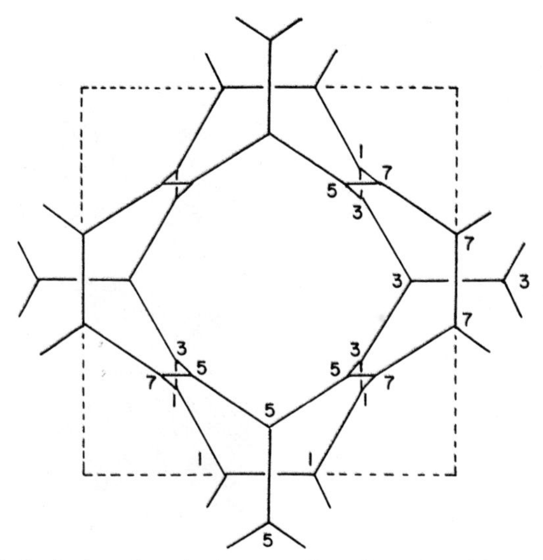

Fig. 5.19. Projection of regular (tetragonal) configuration of the net (9, 3)-*b*.

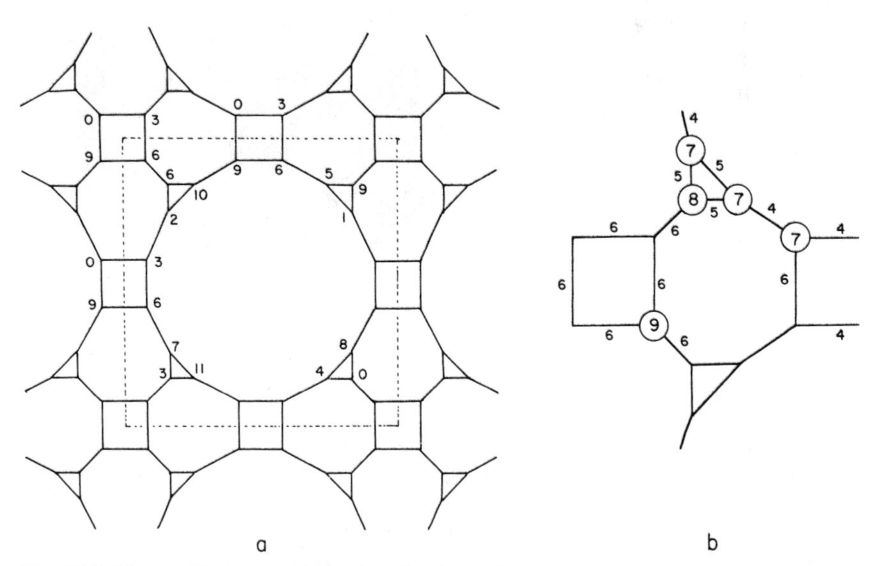

Fig. 5.20. The net (9, 3)-*c*. (*a*) Projection showing heights of points as multiples of $c/12$ for a model constructed from points related by 3_1 and 4_1 axes all of pictch *c*. (*b*) Topological diagram showing *x* and *y* values.

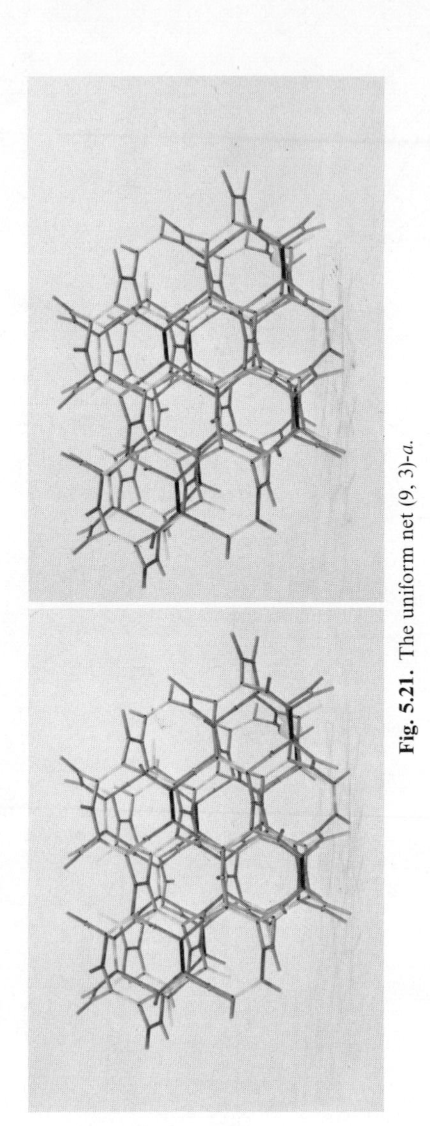

Fig. 5.21. The uniform net (9, 3)-*a*.

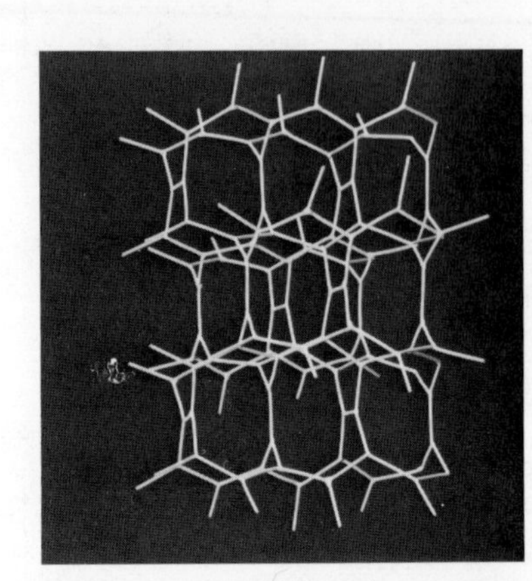

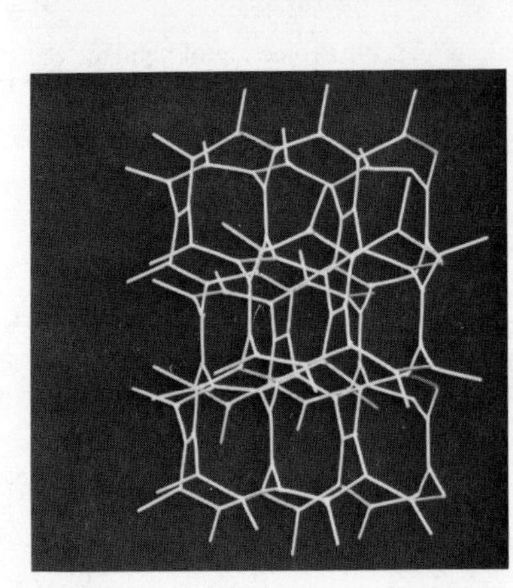

Fig. 5.22. The uniform net (9, 3)-*b*.

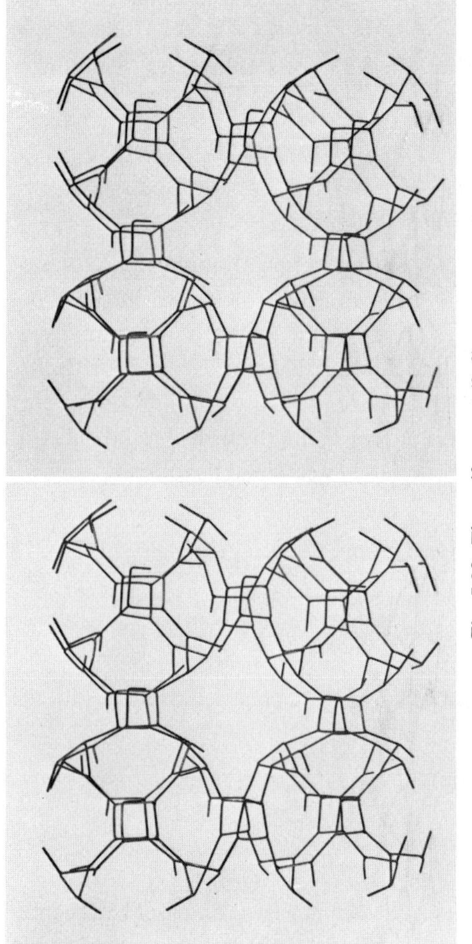

Fig. 5.23. The uniform net (9, 3)-*c*.

different x values ($x_{mean} = 7\frac{1}{2}$) and three different y values ($y_{mean} = 5$) (Fig. 5.20*b*).

Figures 5.21 through 5.23 are stereo-pairs of uniform (9, 3) nets.

Uniform (8, 3) nets

The nets (8, 3)-a and (8, 3)-b

These two nets differ in that in one all the screw axes are 3_1 axes, whereas in the other 3_1 and 3_2 axes alternate, as shown in the projections of Fig. 5.24. The x and y values are the same for both nets, namely, $x = 4$ for all points, $y = 3$ for links in the 3-fold helices, and $y = 2$ for the cross-links (one-third of the total). These nets are illustrated as stereo-pairs in Figs. 5.30 and 5.31 in their most symmetrical configurations (equal links and all bond angles 120°), which are completely described as follows:

(8, 3)-*a* Hexagonal Space group $P6_222$ (No. 180) $Z = 6$

$$c/a = \frac{3\sqrt{2}}{5}; \quad 6(i) \quad x = \tfrac{2}{5}$$

(8, 3)-*b* Rhombohedral Space group $R\bar{3}m$ (No. 166) $Z = 6$ in $(6f)$

Hexagonal cell $c/a = \dfrac{\sqrt{6}}{5}$; $18(f)$ $x = \frac{2}{5}$

The net (8, 3)-c

This net, shown in projection in Fig. 5.25, arises by erecting 2_1 axes perpendicular to the links of the plane 6-gon net. It can be described as a tessellation of 8-gons on the surfaces of a close-packed assembly of parallel hexagonal tunnels. The values of x and y are twice those for a plane 3-connected net. The net is illustrated in its most symmetrical configuration in Fig. 5.32:

Hexagonal Space group $P6_3/mmc$ (No. 194) $Z = 8$

$c/a = \frac{2}{5}$ points in 2(*c*) and 6(*h*) $x = \frac{7}{15}$

All links are of equal length and all bond angles are 120°. Reference was made to the nonequivalence of the links in Chapter 3.

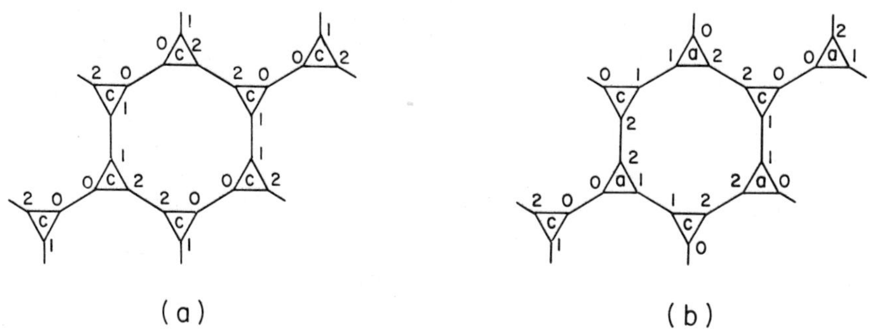

Fig. 5.24. Projections along the 3-fold screw axes of two uniform nets. (*a*) (8, 3)-*a*; (*b*) (8, 3)-*b*.

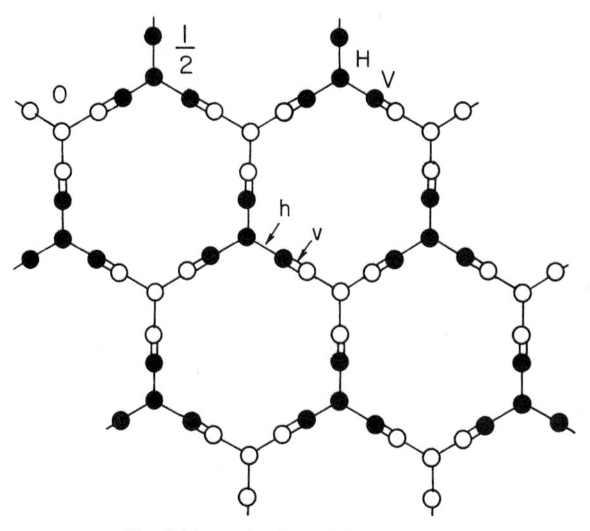

Fig. 5.25. Projection of the net (8, 3)-*c*.

The nets (8, 3)-d, (8, 3)-g, and (8, 3)-h

These three closely related nets arise by erecting 4-fold screw axes at the points and 2-fold screw axes along the links of the plane 4-gon net (Fig. 5.26). They correspond to the similarly related family of (10, 3) nets, *a*, *d*, and *f*, already described. For the points in the 4_1 (4_3) helices, $x = 5$, and for those in the 2_1 helices $x = 3$, giving $x_{\text{mean}} = 4$. For the links in the 4-fold helices, $y = 4$, and for the remainder $y = 2$, giving $y_{\text{mean}} = 2\frac{2}{3}$. In view of the similarity of these three nets, we illustrate only (8, 3)-*d* as a stereo-pair (Fig. 5.33).

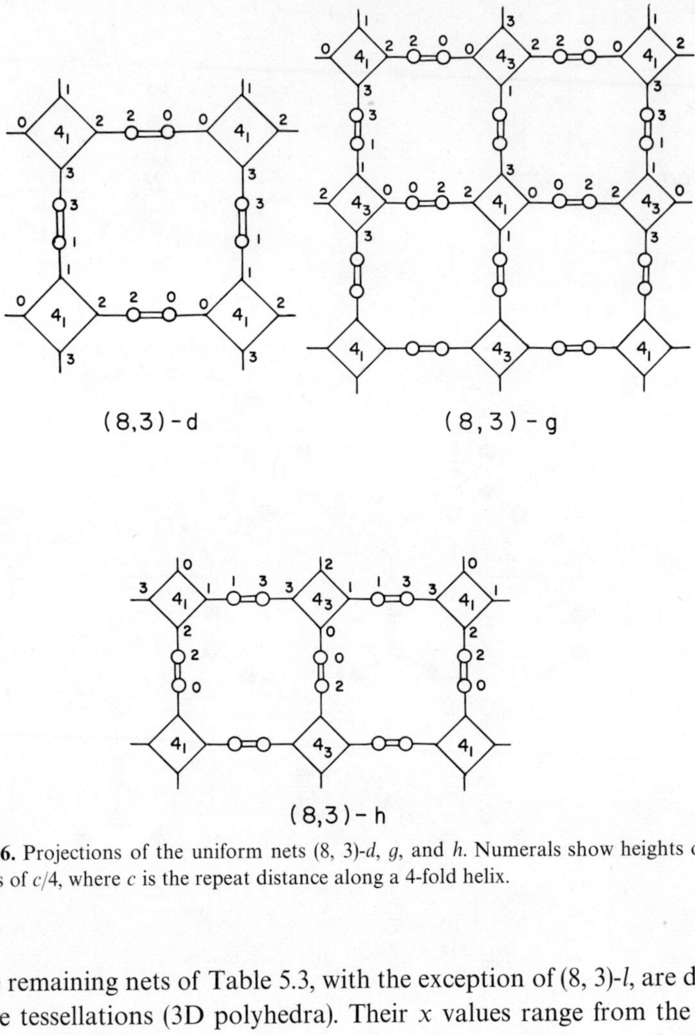

(8,3)-d

(8,3)-g

(8,3)-h

Fig. 5.26. Projections of the uniform nets (8, 3)-*d*, *g*, and *h*. Numerals show heights of points in terms of $c/4$, where c is the repeat distance along a 4-fold helix.

The remaining nets of Table 5.3, with the exception of (8, 3)-*l*, are duals of surface tessellations (3D polyhedra). Their x values range from the lowest possible value (3), which is the value for a 2D tessellation, to the highest value ($7\frac{1}{2}$) found for a (8, 3) net.

The net (8, 3)-e

The 3D polyhedral surface on which this tessellation is inscribed corresponds to the tetragonal net (10, 3)-*b* of Table 5.1. The 3D polyhedron of which this net is the dual is built from the units of Fig. 16.8*a*. This net (not illustrated) is very similar in general appearance to that of Fig. 5.45. Additional 8-gons

Table 5.3 Uniform (8, 3) nets

Net	Figure	Z_t	x	x_{mean}	y	y_{mean}
a*	5.30	6	4	—	2, 3	$2\frac{2}{3}$
b*	5.31	6	4	—	2, 3	$2\frac{2}{3}$
c*	5.32	8	6	—	4	—
d	5.33	8	3, 5	4	2, 4	$2\frac{2}{3}$
e	—	8	4, 5	$4\frac{1}{2}$	2, 3, 4	3
f	5.34	8	6, 7, 8, 9	$7\frac{1}{2}$	3, 4, 5, 6, 7	5
g	5.26	16	3, 5	4	2, 4	$2\frac{2}{3}$
h	5.26	16	3, 5	4	2, 4	$2\frac{2}{3}$
i†	5.35	16	3	—	2	—
j	5.36	16	4	—	2, 3	$2\frac{2}{3}$
k	5.37	24	3	—	2	—
l	5.38	24	4, 5	$4\frac{2}{3}$	2, 3, 4	$3\frac{1}{9}$
m	5.39	32	3, 4	$3\frac{3}{4}$	2, 3	$2\frac{1}{2}$
n*	5.40	16‡	3	—	2	—
o	5.41	48	4, 5, 6	5	2, 3, 4, 5	$3\frac{1}{3}$

* Net has regular configuration (equal bonds, bond angles 120°).
† This net was described in Part 7 as the dual of the polyhedron of Fig. 16.11 (present work), which is based on a distorted diamond net. Further study of a large model would be desirable to check the repetition of this unit in three dimensions.
‡ Both the nets m and n are duals of 6-tunnel 3D polyhedra, but the value of Z_t is halved in the latter case because the polyhedron (and therefore the dual net) is body-centered.

around the tunnels result in equal numbers of points having $x = 4$ and 5, whence $x_{mean} = 4\frac{1}{2}$. The y values for links meeting at the two kinds of point are

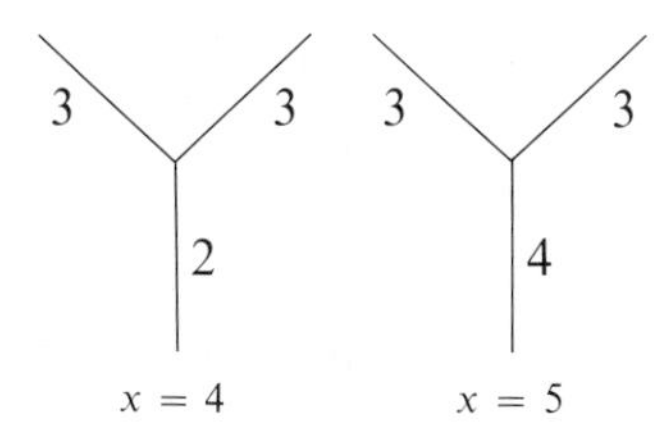

The net (8, 3)-f

This net (Fig. 5.34) is closely related to (8, 3)-e, being derived from the cubic net (10, 3)-a by inflating the links and inscribing the tessellation on the surface so formed. The 3D polyhedron (3, 8) of which this net is the dual is illustrated in Fig. 16.9. In contrast to the net (8, 3)-*e*, there are numerous 8-gons around the tunnels in addition to those on the surface, and from the topological standpoint this is the most complex net yet studied. Of the eight points in the topological repeat unit, one has $x = 6$, three have $x = 7$, three have $x = 8$, and one has $x = 9$, giving $7\frac{1}{2}$ as the weighted mean. The y values associated with the four kinds of nonequivalent point are as follows:

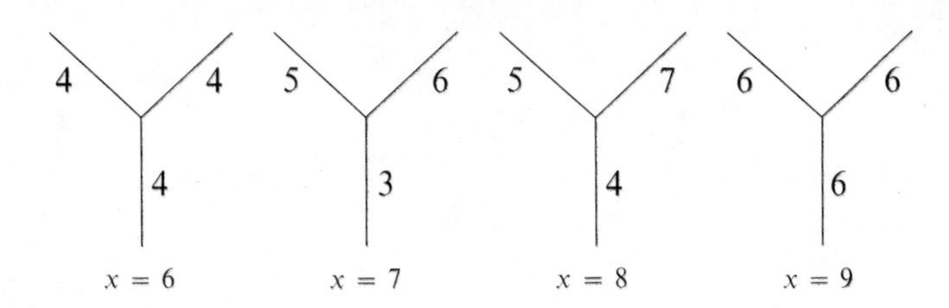

The net (8, 3)-i

The nets (8, 3)-*i*, *k*, and *n* are of special interest as having values of x and y the same as for a plane tessellation, namely, 3 and 2. In the duals of the 3D polyhedra this arises if there are no 8-gon circuits around the tunnels. In this net there are numerous 9-gons but no 8-gons other than those on the surfaces of the tunnels (Fig. 5.35).

The net (8, 3)-j

This net is derived from a 3D polyhedron (Fig. 16.15) in which four coplanar tunnels meet at each point of the underlying net. In the derived (8, 3) net all points have $x = 4$. The value of Z in Table 5.3 corresponds to the dual of the (3, 8)-4*t* polyhedron. The net is illustrated in Fig. 5.36.

The net (8, 3)-k

This net (Fig. 5.37), not previously described, is the dual of the trigonal (3 + 2)*t* polyhedron of Fig. 16.16 (or of the complementary (6 + 2)*t* polyhedron). It is shown in projection in Fig. 5.27. It cannot be built with equal coplanar bonds inclined at 120°; that is, it is not a *regular* net. There are three types of nonequivalent point (all with $x = 3$) and five sets of nonequivalent

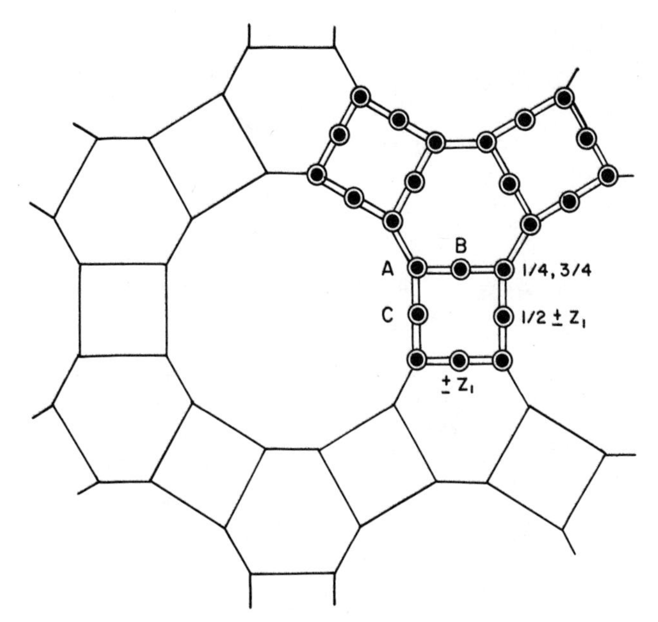

Fig. 5.27. Projection of the net (8, 3)-*k*.

links (all with $y = 2$). The value of $Z = 24$ in Table 5.3 corresponds to the dual of the 5*t* polyhedron (for which $Z = 9$).

The net (8, 3)-l

This net has not been described in earlier papers. The portion forming the model of Fig. 5.38 is shown in projection in Fig. 5.28, where the broken lines are edges of the b.c. tetragonal cell and the heavy lines represent planes of symmetry. There are two kinds of nonequivalent point, *A* and *B*, and four kinds of nonequivalent link. The topological data are summarized below, where the numbers of points and links are those in the b.c. tetragonal cell ($Z = 48$):

Point	x	Number per cell
A	4	16
B	5	32

Link	y	Number per cell
a	3	32
b	2	8
c	3	16
d	4	16

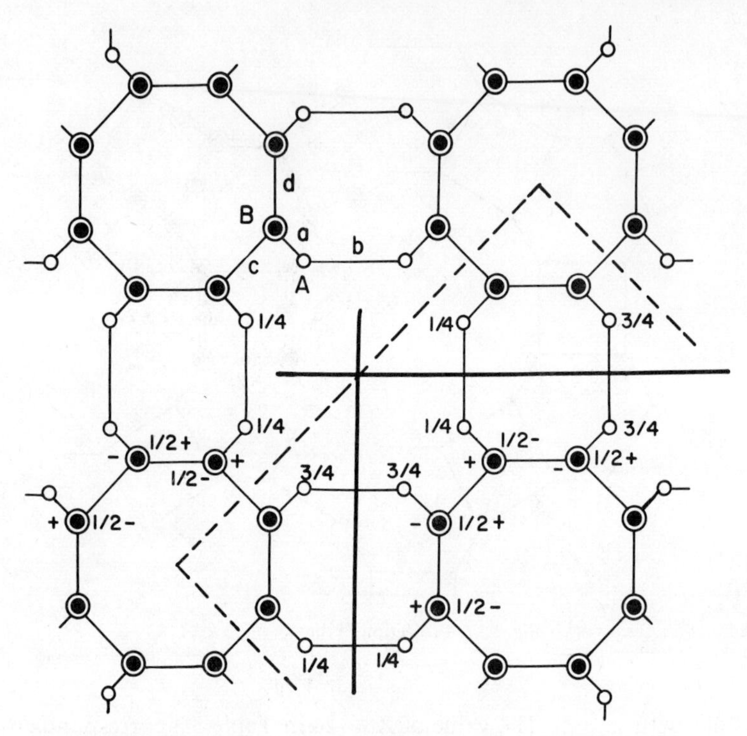

Fig. 5.28. Projection of the tetragonal configuration of the net (8, 3)-*l*.

The net (8, 3)-m

This net (Fig. 5.39) is the dual of the 6-tunnel 3D polyhedron (3, 8) of which a repeat unit is illustrated in Fig. 16.8*b*. In this net eight points in the repeat unit have $x = 3$ and the other 24 have $x = 4$, giving $x_{\text{mean}} = 3\frac{3}{4}$. Half the links have $y = 2$ and the remainder have $y = 3$.

The net (8, 3)-n

This net, illustrated stereoscopically in Fig. 5.40, is the dual of the polyhedron of Fig. 16.8*c*. As in the net (8, 3)-*i* there are no 8-gons other than those on the surface of the original polyhedron, and therefore $y = 2$. This net is one of the relatively small number of 3D 3-connected nets that can be constructed with equal links and bond angles of 120°. Figure 5.29 is a projection along the *c* axis of the most symmetrical configuration of this net. The net is completely

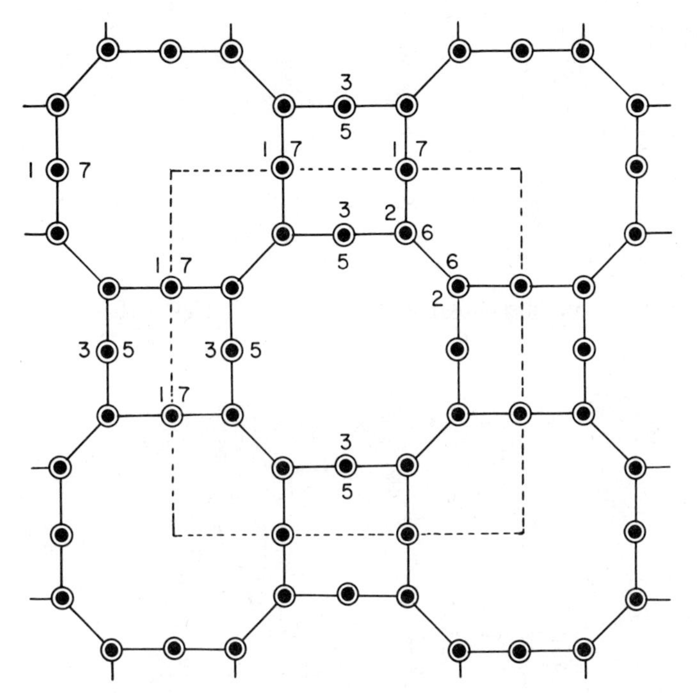

Fig. 5.29. Projection on (001) of the regular configuration of the net (8, 3)-*n*. Numerals indicate heights of points in terms of $c/8$.

described as follows:

Tetragonal Space group $I4/mmm$ (No. 139) $Z = 32$

$$c/a = \frac{4}{2\sqrt{3} + \sqrt{2}} = 0.820$$

$16(k)$: $(x, \frac{1}{2} + x, \frac{1}{4})$; $x = \dfrac{\sqrt{3} + \sqrt{2}}{2(2\sqrt{3} + \sqrt{2})} = 0.322$

$16(n)$: $(0xz)$; $x = 0.322$; $z = \frac{1}{8}$

The net (8, 3)-o

This net is the dual of the 8-tunnel 3D polyhedron built from the repeat unit shown in Fig. 16.17. Of the 48 points in the topological repeat unit, 8 have $x = 4$, 32 have $x = 5$, and 8 have $x = 6$, giving a weighted mean value of x

equal to 5:

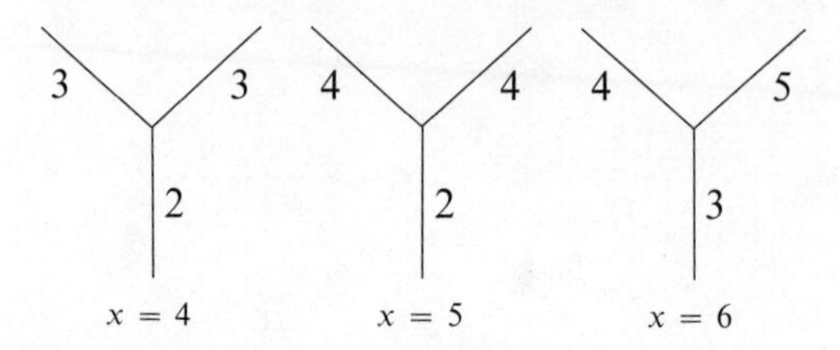

Two units of the net are shown as a stereo-pair in Fig. 5.41.

Figures 5.30 through 5.41 are stereo-pairs of uniform (8, 3) nets.

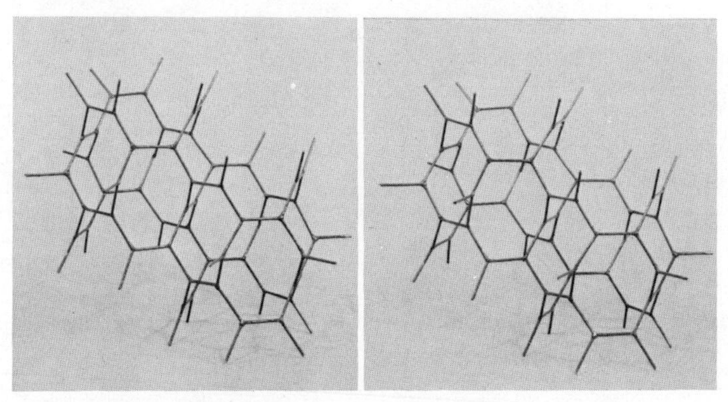

Fig. 5.30. The uniform net (8, 3)-*a*.

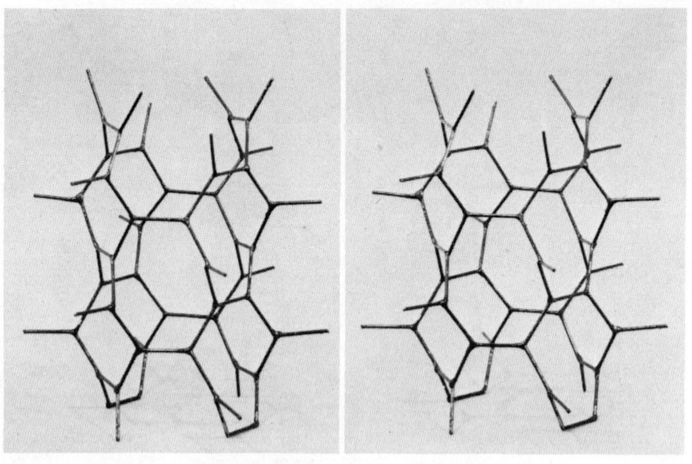

Fig. 5.31. The uniform net (8, 3)-*b*.

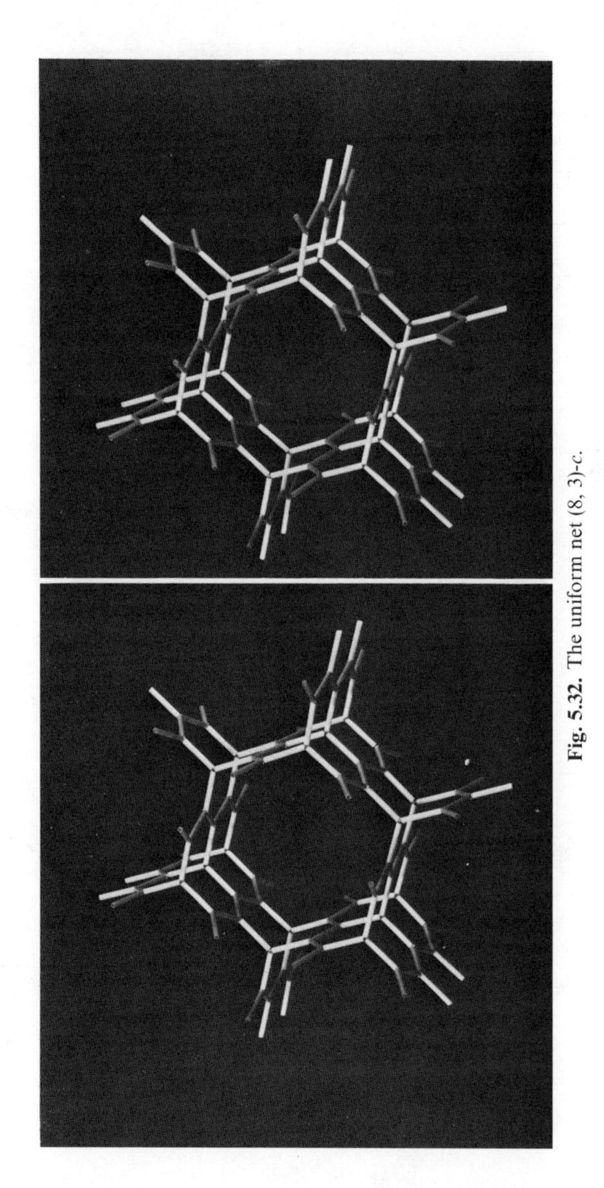

Fig. 5.32. The uniform net (8, 3)-*c*.

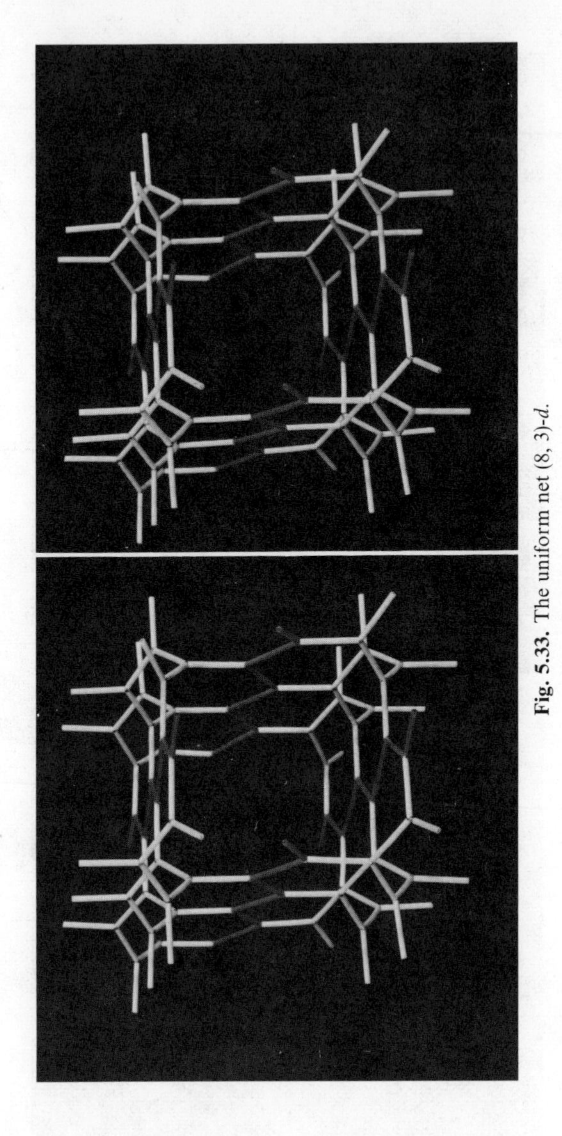

Fig. 5.33. The uniform net (8, 3)-*d*.

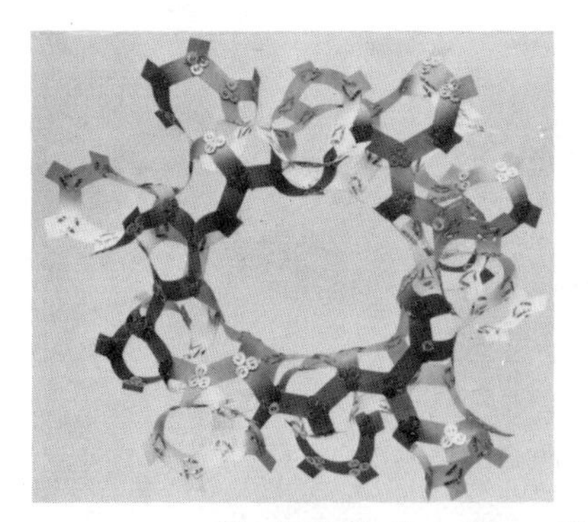
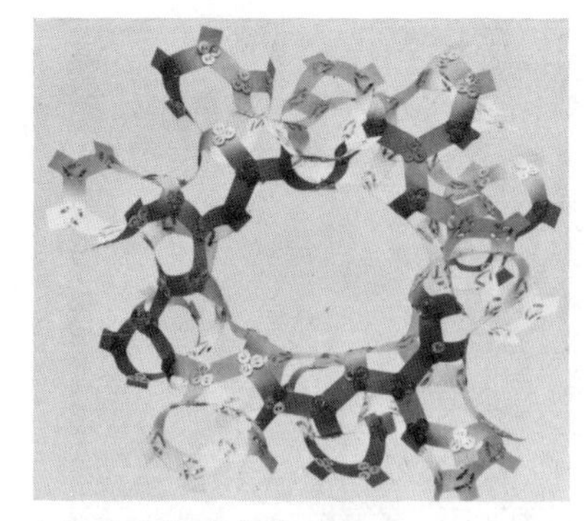

Fig. 5.34. Portion of the uniform net (8, 3)-*f*, dual of Fig. 16.9.

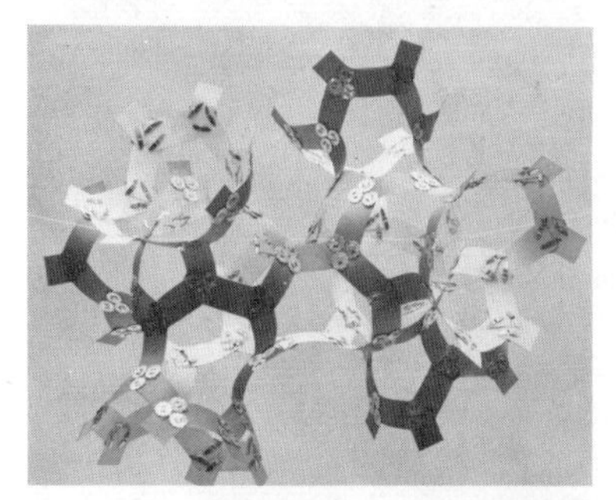
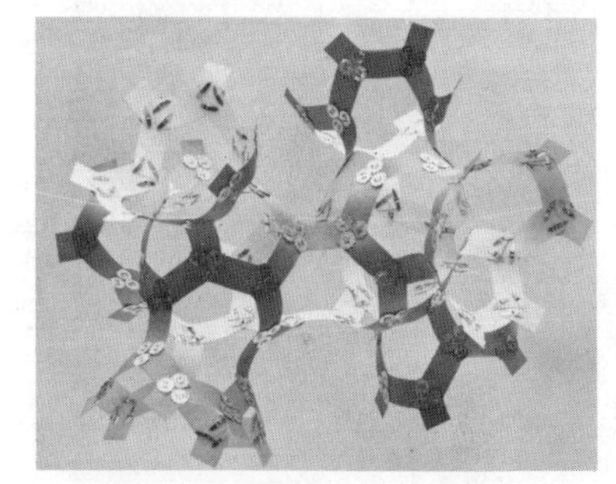

Fig. 5.35. Portion of the uniform net (8, 3)-*i*, dual of Fig. 16.11.

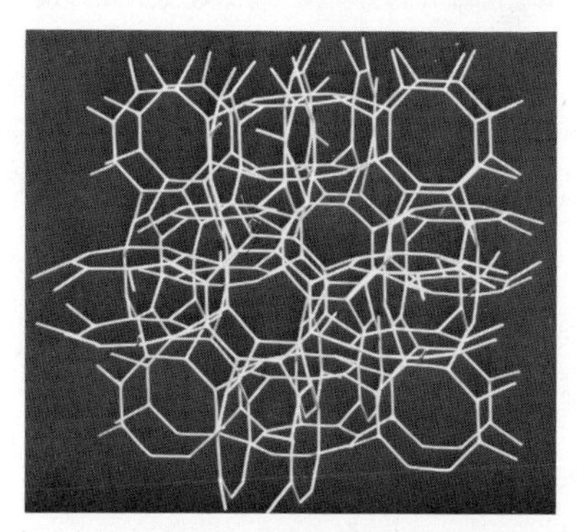
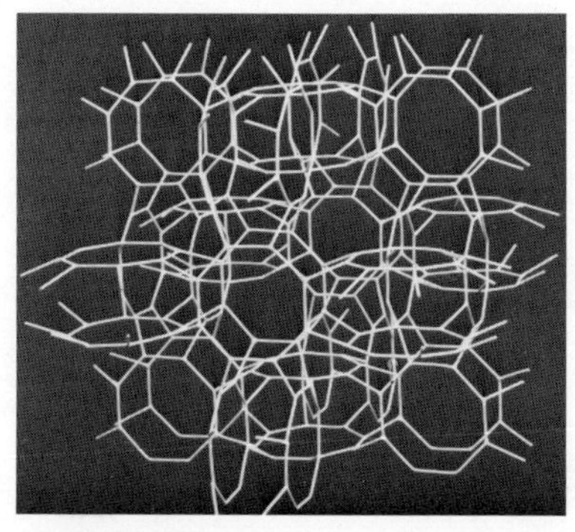

Fig. 5.36. The uniform net (8, 3)-*j*, dual of Fig. 16.15.

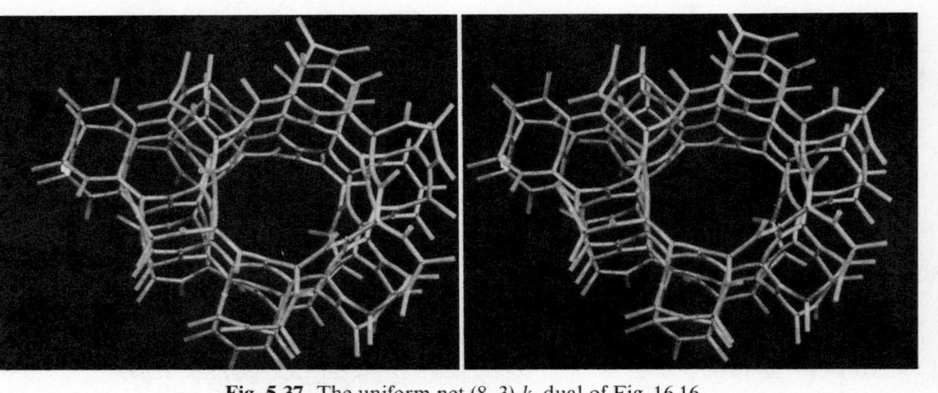

Fig. 5.37. The uniform net (8, 3)-*k*, dual of Fig. 16.16.

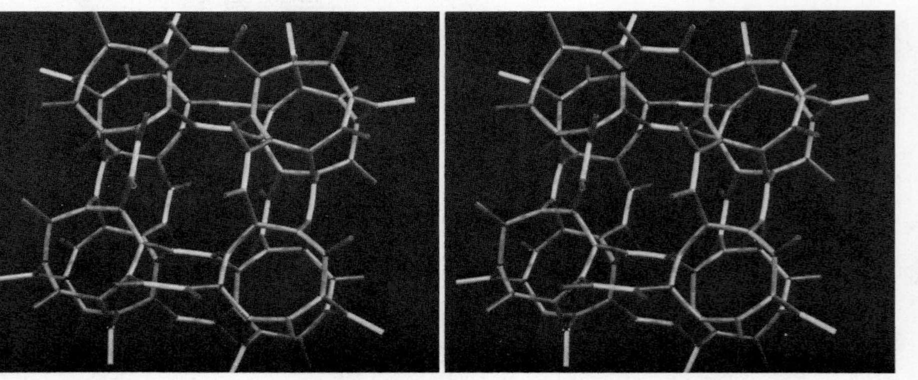

Fig. 5.38. The uniform net (8, 3)-*l*.

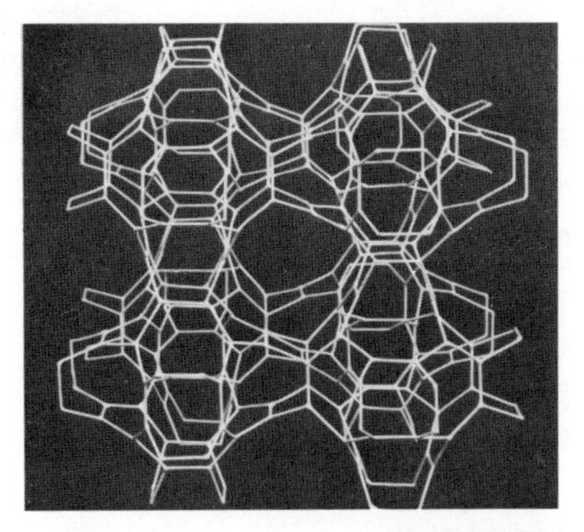
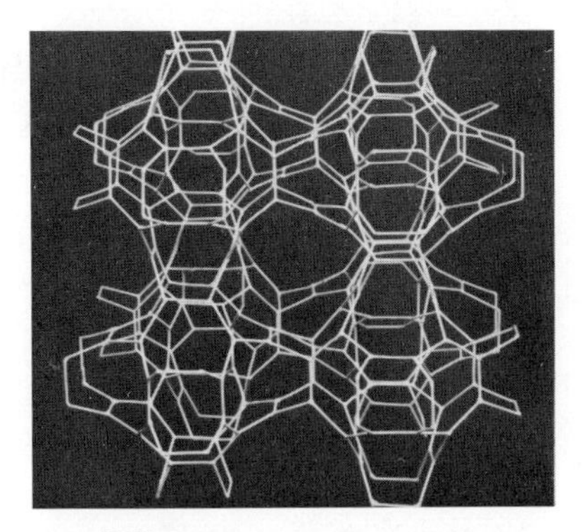

Fig. 5.39. The uniform net (8, 3)-*m*.

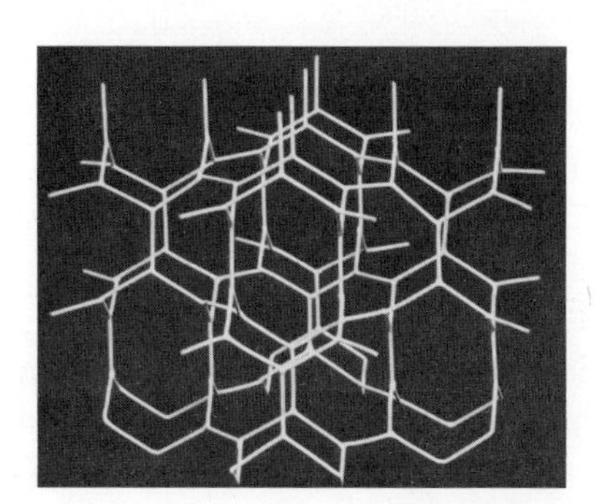
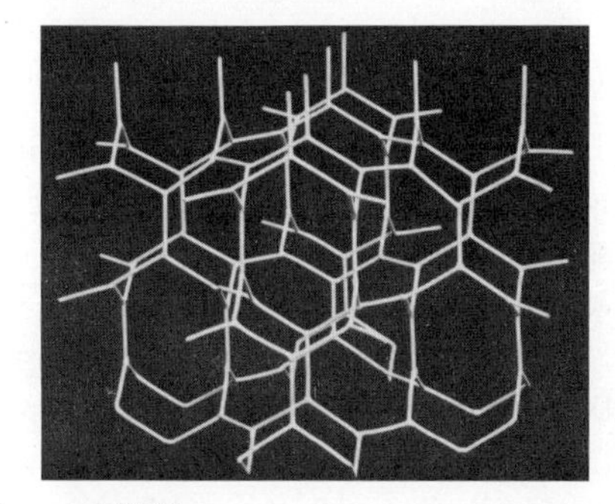

Fig. 5.40. The uniform net (8, 3)-*n*.

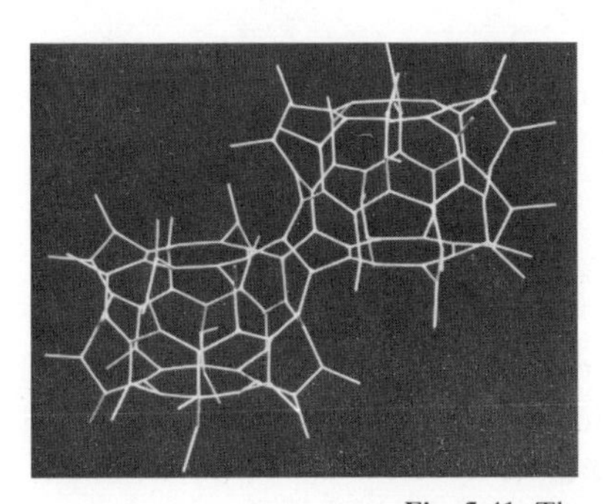
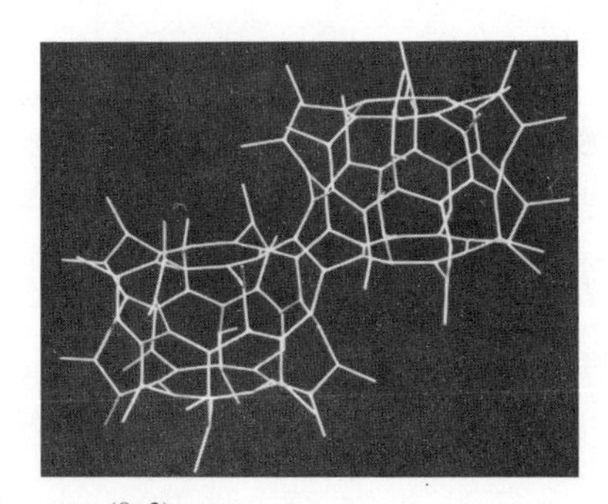

Fig. 5.41. The uniform net (8, 3)-*o*.

Uniform (7, 3) nets

The four known uniform (7, 3) nets are listed in Table 5.4.

Table 5.4 Uniform (7, 3) nets

Net	Z_t	x	x_{mean}	y	y_{mean}
(7, 3)-*a* } (7, 3)-*b* }	12	3 (six)	$\frac{7}{2}$	2 (six)	$\frac{7}{3}$
		4 (six)		2 (six)	
				3 (six)	
(7, 3)-*c*	14	3		2	
(7, 3)-*d*	84	3 (36)	$\frac{11}{3}$	2 (74)	$\frac{22}{9}$
		4 (40)		3 (48)	
		5 (8)		4 (4)	

The nets (7, 3)-a and (7, 3)-b

These two nets are related in a similar way to the pair (8, 3)-*a* and (8, 3)-*b*, as is evident from Fig. 5.42. In the net (7, 3)-*a* all the 3-fold helices are of the same kind; in the net (7, 3)-*b* 3_1 and 3_2 axes alternate. There are equal numbers of points in the 2_1 and $3_1(3_2)$ helices, with $x = 3$ and 4, respectively, giving $x_{mean} = 3\frac{1}{2}$. There are equal numbers of links of three kinds, namely, those in the 2-fold helices ($y = 2$), those connecting the helices ($y = 2$), and those in the 3-fold helices ($y = 3$), giving $y_{mean} = \frac{7}{3}$. The models illustrated stereoscopically in Figs. 5.43 and 5.44 are constructed with coplanar bonds at each point, necessitating longer links in the 2-fold helices.

The net (7, 3)-c

This net is the dual of the 3D polyhedron (3, 7) of Fig. 16.4, which is a tessellation of triangles on the surface formed by inflating the links of the net (10, 3)-*b*. Since the smallest circuit around a tunnel is an 8-gon, x has the same value (3) as for a plane 3-connected net. A model of the net is presented in Fig. 5.45.

The net (7, 3)-d

This net is the dual of the (3, 7)-8*t* polyhedron of which one unit is illustrated in Fig. 16.7. Being the dual of an 8-tunnel polyhedron for which $Z = 36$, it has 84 points in the (topological) repeat unit. The unit has *mmm* symmetry and forms a pseudo-body-centered structure in which adjacent units are

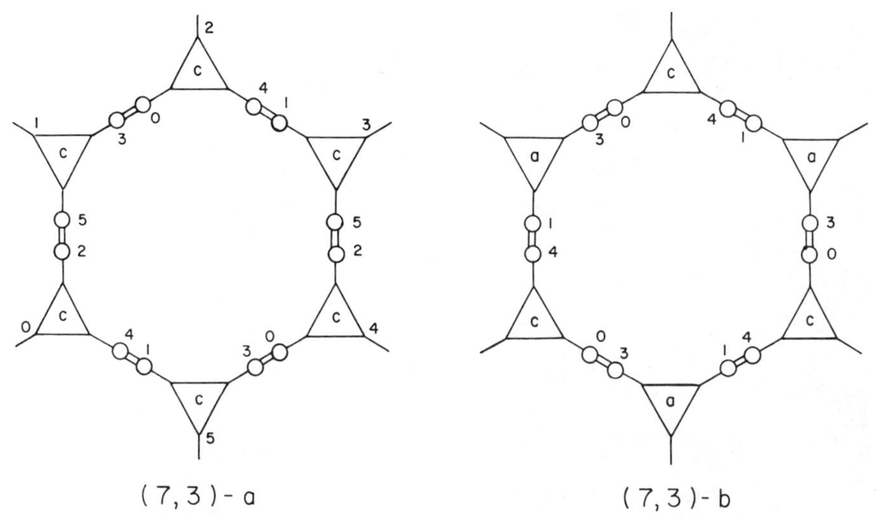

Fig. 5.42. Projections of the uniform nets (7, 3)-*a* and *b*. Numerals indicate heights of points as multiples of $c/6$; *c* and *a* indicate 3_1 and 3_2 axes.

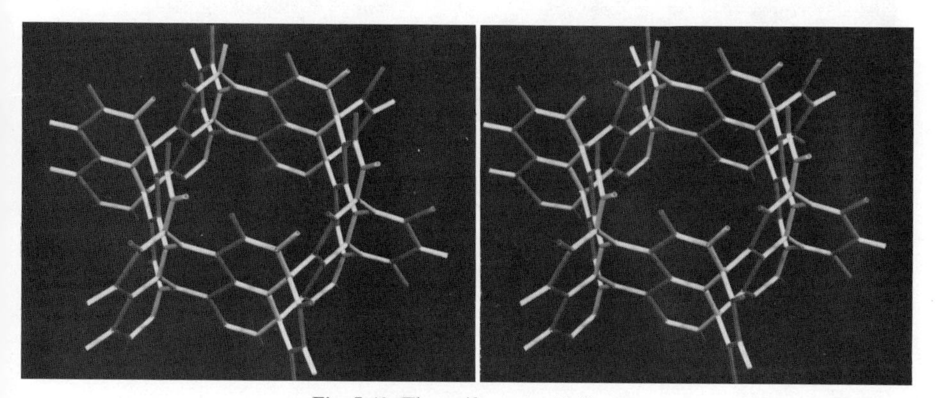

Fig. 5.43. The uniform net (7, 3)-*a*.

related by a rotation through 90°. A model of one unit is presented in Fig. 5.46. Only the points and links involved in the two 7-gon circuits around each tunnel have x and y values different from those of a plane net (3 and 2, respectively). Of the 84 points, 36 have $x = 3$, 40 have $x = 4$, and 8 have $x = 5$. Of the 126 links, 74 have $y = 2$, 48 have $y = 3$, and 4 have $y = 4$. The weighted mean values of x and y are $\frac{11}{3}$ and $\frac{22}{9}$, respectively.

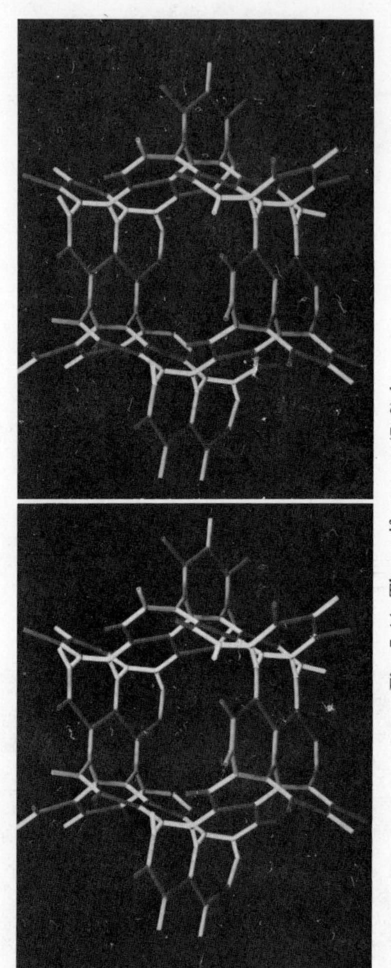

Fig. 5.44. The uniform net (7, 3)-*b*.

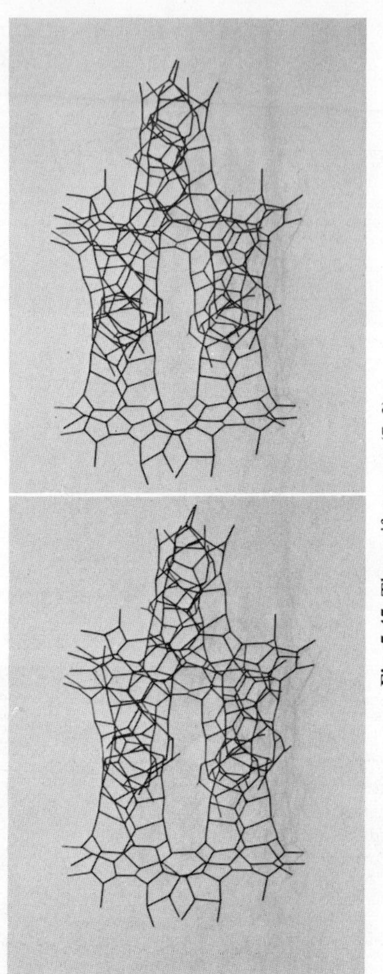

Fig. 5.45. The uniform net (7, 3)-*c*.

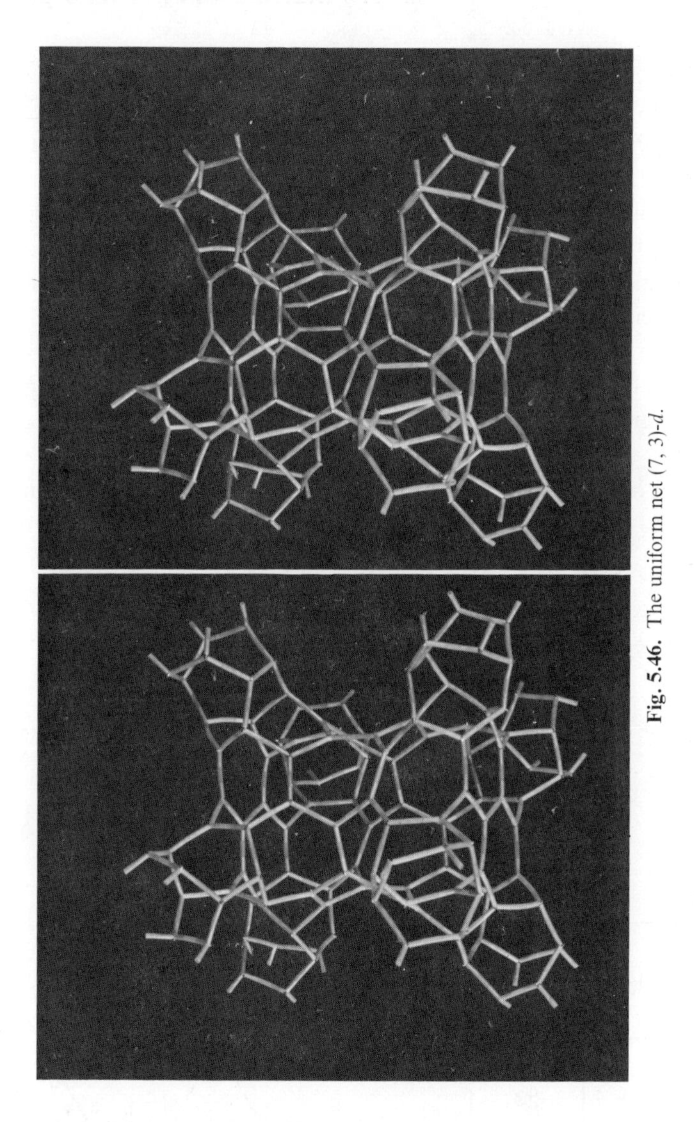

Fig. 5.46. The uniform net (7, 3)-*d*.

Appendix: relation between the nets (10, 3)-*b* and (10, 3)-*c*

In the nets (10, 3)-*b* and (10, 3)-*c* the characteristic subunit is the planar zigzag chain. Repetition of such a chain by 2_1 axes in the plane of the chains gives the planar 6-gon net. In (10, 3)-*b* the chain directions are the *a* and *b* axes, and the chains are repeated around 4_1 and 4_3 axes that are parallel to the *c* axis. In (10, 3)-*c* the chains are parallel to three equivalent directions perpendicular to *c* and they are repeated by the operation of 3_1 axes. Owing to the special values of the coordinates, namely, $x = \frac{1}{3}$, $y = \frac{1}{6}$, and $z = \frac{1}{9}$ in the general position 6(*c*) of $P3_112$, there are also 6_4 axes as shown in Fig. 5.47. Since nets can be generated from zigzag chains by the operation of 2_1, 3_1, or 4_1 axes, it is of interest to show that there is no analogous net generated by 6_1 axes.

In a net of this family generated by screw axes n_1 in which all interbond angles are 120° and all links equal in length with one-third of them parallel

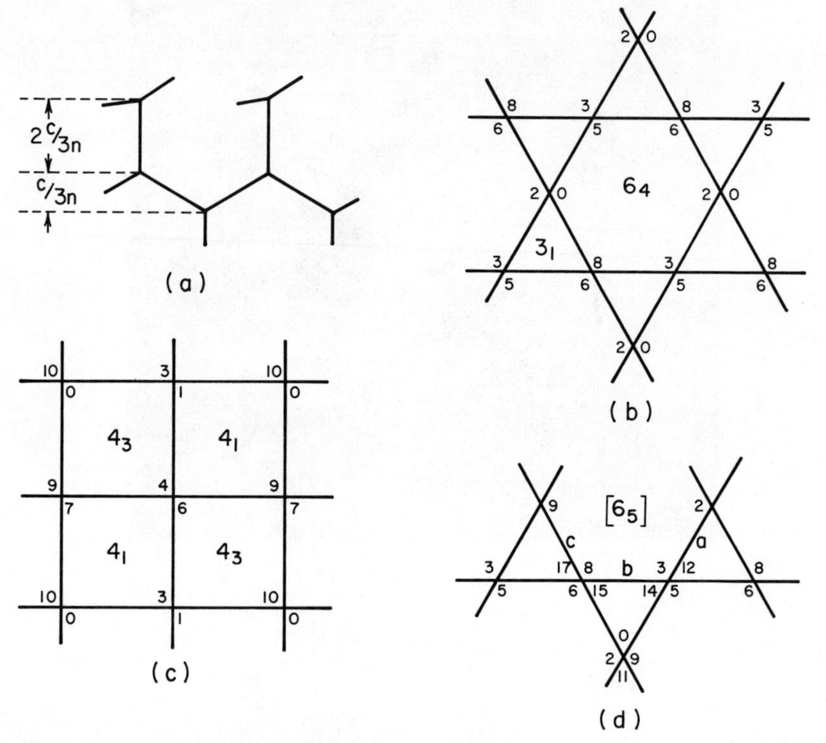

Fig. 5.47. (*a*) Dimensions of zigzag chain; arrow indicates direction of the screw axis. (*b*) Projection of (10, 3)-*c* (unit of height *c*/9). (*c*) Projection of (10, 3)-*b* (unit of height *c*/12). (*d*) Zigzag chains related by 6_1 axis (unit of height *c*/18).

to the screw axes, the dimensions of the zigzag chain are those of Fig. 5.47*a*. The numbers in Fig. 5.47*b*, *c* and *d* are therefore heights of points in terms of $c/9$, $c/12$, and $c/18$, respectively, for these diagrams are projections along the screw axes. Each point of intersection of two chains in these projections must correspond to a pair of points in the net which are separated in height by two units ($c/9$, etc.), this being the length of the (vertical) link cross-linking the chains. To facilitate comparison with the projection of (10, 3)-*c* in Fig. 5.47*b*, a 6_5 is used instead of a 6_1 axis. In Fig. 5.47*d* successive chains are related by a 6_5 axis, starting with chain *a* (points at heights 2/18 and 3/18). This is converted into chain *b* (heights 5/18, 6/18) then into *c* (heights 8/18, 9/18), and so on. Above the points of intersection of chains *a* and *c* in the projection there are points at heights (2/18 and 9/18) which cannot be joined by a link of length $2c/18$. It might appear that this difficulty could be overcome by introducing further points such as that at 0/18, which would then belong to a chain vertically below the chain *c* and one at 11/18, which would belong to a chain vertically above chain *a*. These in turn would require points as indicated, at heights 17/18, 15/18, 14/18, and 12/18. The final result is the production of four net points above each point of the projection, these consisting of pairs separated by 9/18. The repeat distance along the *c* axis is therefore halved, and the heights may be read as multiples of one-ninth of the new *c* axis. Evidently the net is identical with (10, 3)-*c*, whose 6-fold screw axis is not 6_5 but 6_4. There is accordingly no net corresponding to (10, 3)-*b* and (10, 3)-*c* but generated by 6_1 (or 6_5) axes.

6

Some Archimedean 3-connected nets

No attempt has been made to derive systematically 3-connected nets to which we give the name "Archimedean"; that is, nets in which the shortest circuits from each point are two polygons of one kind and one of another, or three of different kinds, the nature of the polygons being the same for each point (point symbol $n_a n_b^2$ or $n_a n_b n_c$). The nets described in this section have been found during studies of various aspects of 3-connected nets. Each net can be built with all links of equal length and coplanar at each point, but only three can have configurations with equal bonds *and* equal interbond angles (120°):

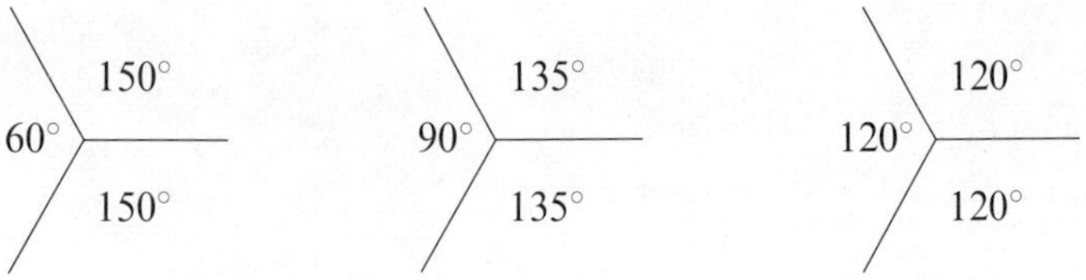

60° | 150° | 150° — 3.20²-cubic

90° | 135° | 135° — 4.12²-cubic

120° | 120° | 120° — 6.8²-cubic, 8².10-*a*, 8².10-*b*

← 4.14² →

← 6.10² →

The net 4.14^2 can have one angle in the range 60°–90° and 6.10^2 can have one angle in the range 60°–120°, but in neither case can the angle attain one of the extreme values. We consider first the group of three cubic nets, all having 24 points in the unit cell. In their most symmetrical configurations they are described as follows:

Net	Space group	Equivalent positions	Angles between links (number)
3.20^2	$I4_13$	$24(h)$: $x = 0.058$	60° (1), 150° (2)
4.12^2	$Im3m$	$24(g)$: $x = 0.147$	90° (1), 135° (2)
6.8^2	$P\bar{4}3m$	$24(j)$: $x = \frac{5}{12}$, $y = \frac{3}{12}$, $z = \frac{1}{12}$	120° (3)

The net 3.20^2

The cubic configuration of this net results from replacing the points in the cubic configuration of (10.3) by equilateral triangles; it corresponds to the open sphere packing 3_2 of Heesch and Laves. Since the shortest circuits in the original net are 10-gons, the new net consists of 3-gons and 20-gons. Nets of the same kind are derivable from other uniform 3-connected nets, which give rise to nets consisting of 3-gons and, respectively, 14-, 16-, 18-, 20, and 24-gons. These nets are obviously related to the Archimedean semiregular solids that result from truncating the three regular solids of which three edges meet at each vertex, namely, the tetrahedron, cube, and regular dodecahedron. The connecting link between these finite polyhedra and the 3D nets is the plane net 3.12^2 in which triangles replace the single points of the plane 6-gon net (Fig. 6.1).

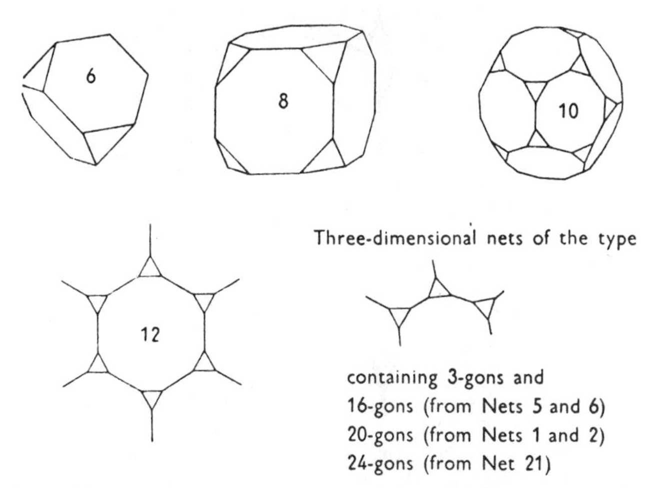

Fig. 6.1. The semiregular solids, plane net, and 3D nets forming the family $\{3, n^2\}$. This figure is reproduced from the original paper (Part 5) and the serial numbers of the nets are those of Table 4.2. Nets 3.14^2 and 3.18^2 should now be included.

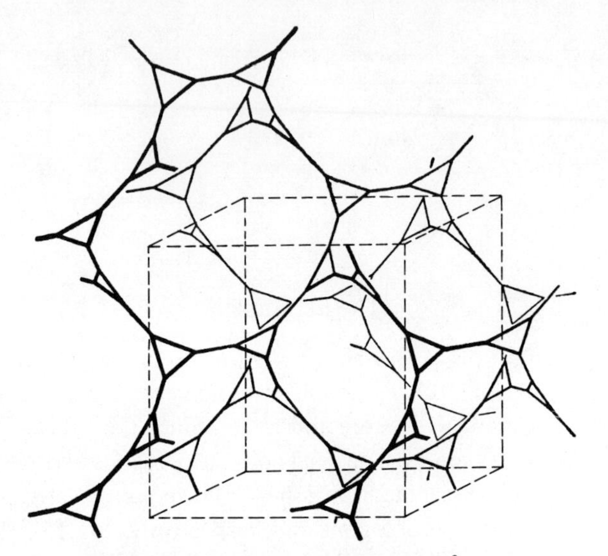

Fig. 6.2. The 3-connected net 3.20^2.

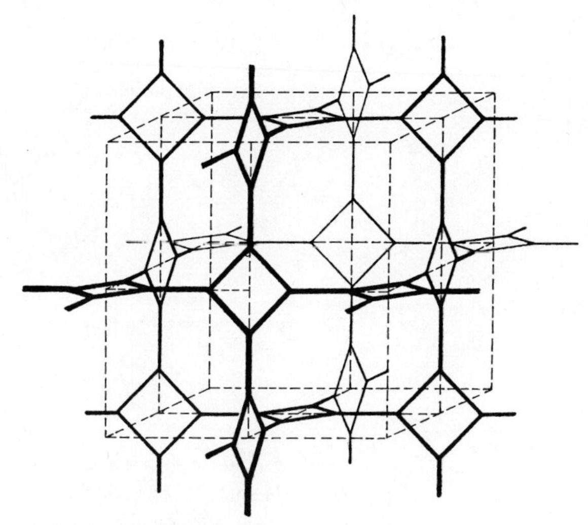

Fig. 6.3. The 3-connected net 4.12^2.

The net 3.20^2 (Fig. 6.2) is apparently the only net of this family having a configuration with cubic symmetry. The remaining nets noted in Fig. 6.1 have not been studied in detail.

The net 4.12^2

Replacement of the points of the NbO net by 4-gons gives the net 4.12^2 illustrated in Fig. 6.3. We noted earlier that the 3-connected nets (10, 3)-*a* and (10, 3)-*b* result from replacing the points in the diamond net by the unit

which can be oriented in two different ways. The same process may be applied to the net of Fig. 9.21*a*, which corresponds to the edges of the polyhedra in Fedorov's space-filling arrangement of truncated octahedra. If the unit is inserted at each point so that the line joining the two substituting points bisects the pair of opposite 90° bond angles, the net 4.12^2 is obtained.

The net 6.8^2

This net, the only one of this group that can be built with equal bonds and equal interbond angles, arises by replacing the points in the Fedorov net by two-point units oriented so that the lines joining the pairs of points bisect a pair of opposite 120° bond angles. This net has tetrahedral symmetry, the planes of the hexagons being parallel to the faces of a tetrahedron (Fig. 6.4).

The nets $8^2.10$-*a* and $8^2.10$-*b*

We have seen that three uniform (10, 3) nets project as the same 4.8^2 planar net, namely, (10, 3)-*a*, (10, 3)-*d*, and (10, 3)-*f*. By joining the helices at different levels other nets are formed that have the same projection but contain 4-gon or 8-gon circuits (Fig. 6.5). Some of these are Archimedean nets; that is, all points are of the type $n_1n_2^2$ (or $n_1^2n_2$). Of these, $8^2.10$-*a* is illustrated in Fig. 6.6, and 4.14^2 is shown as a stereo-pair in Fig. 6.9.

Owing to the similarity of these nets $8^2.10$, we illustrate only $8^2.10$-*b* as a stereo-pair (Fig. 6.7) and we show two more detailed projections of this net in Fig. 6.8. Both these $8^2.10$ nets can be constructed with equal bonds and interbond angles of 120°.

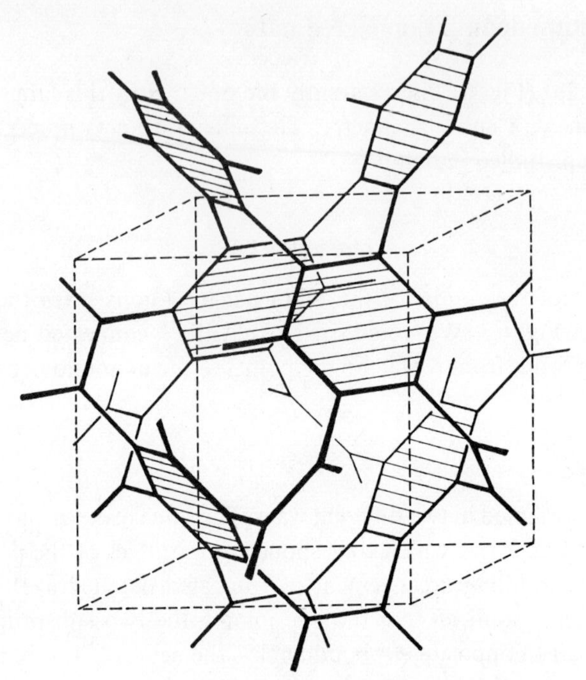

Fig. 6.4. The 3-connected net 6.8^2.

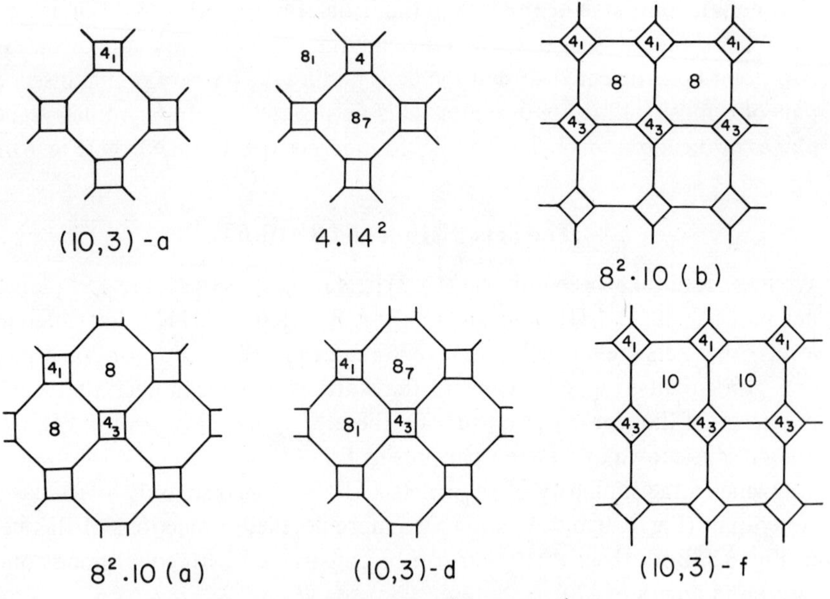

Fig. 6.5. A family of related 3-connected nets.

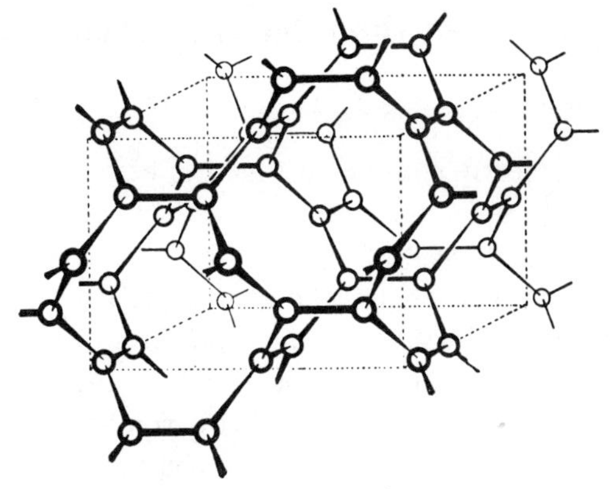

Fig. 6.6. The net $8^2.10$-*a*.

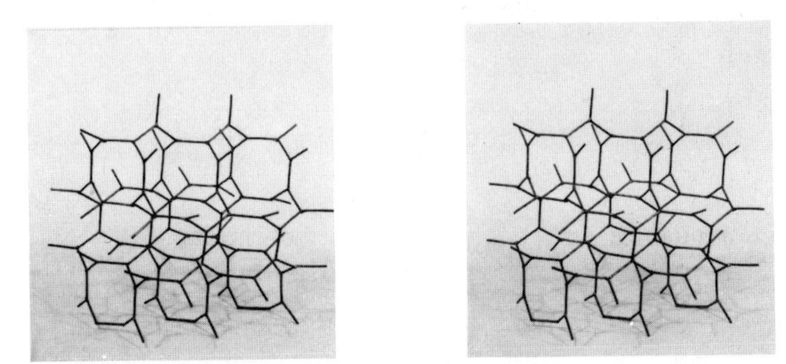

Fig. 6.7. The net $8^2.10$-*b*.

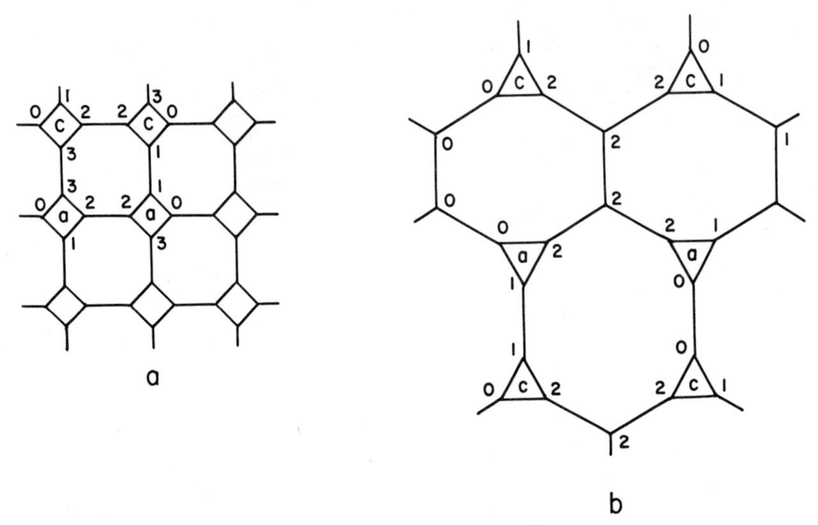

Fig. 6.8. Projections of the net $8^2.10$-*b*.

The nets 4.14^2 and 6.10^2

These nets are conveniently considered together because they are derived in a similar way from simpler nets and because they have somewhat similar geometrical properties. The net 4.14^2 is one of those that project as the planar 4.8^2 net (Fig. 6.5), and one configuration of this net arises by replacing the points of the diamond net by squares whose planes are all parallel to (001). This form of the net is illustrated stereoscopically in Fig. 6.9. All the bonds are equal in length, but the bonds meeting at a point are not coplanar. At first sight the latter feature might appear to be characteristic of this net, but its geometry is variable within wide limits. The angle α (Fig. 6.10) is variable, still retaining the general configuration of Fig. 6.9 with tetragonal symmetry:

α	0°	45°	60°	90°
c/a	0	0.667	0.905	1.414
x	0.146	0.167	0.185	0.250

The x coordinate refers to a b.-c. tetragonal cell of the diamond net in which the point A in Fig. 6.10 would be (x00) referred to the origin O. The lower

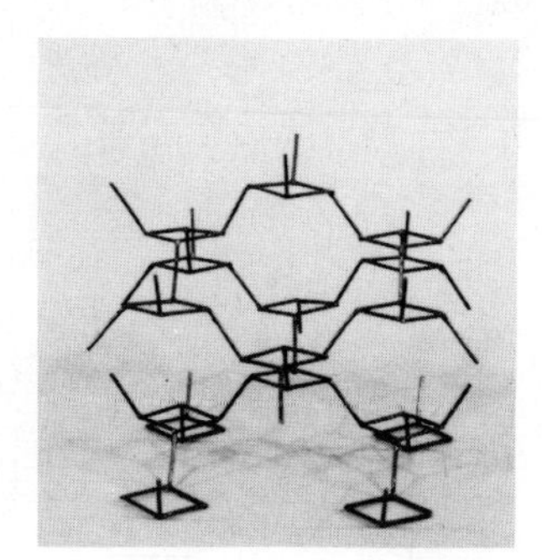
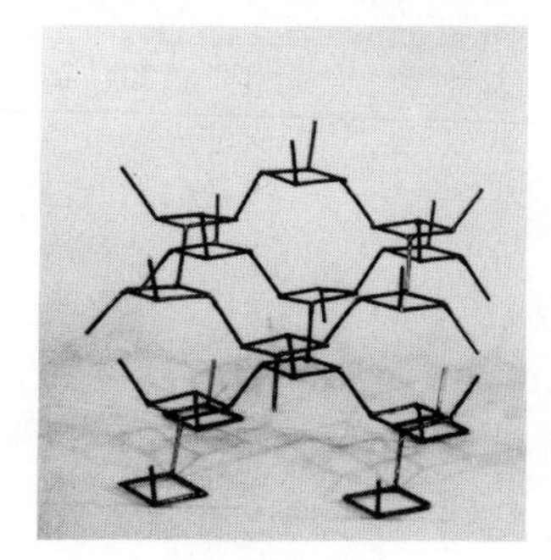

Fig. 6.9. The net 4.14^2.

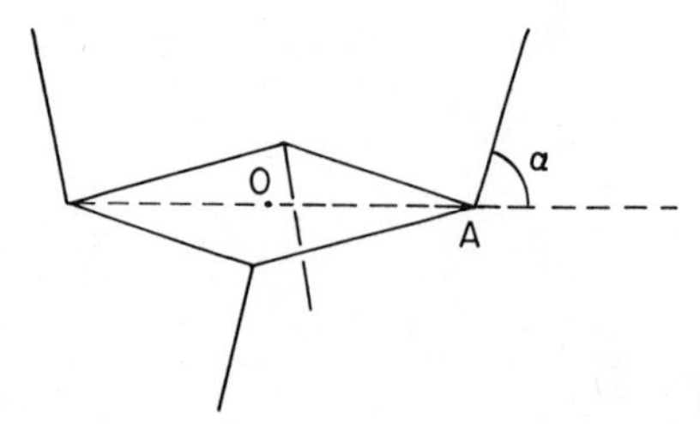

Fig. 6.10. Geometry of the net 4.14^2 (see text).

limiting value of the axial ratio corresponds to compression of the whole 3D net into the planar 4.8^2 net. More interesting configurations arise if the four points around each node of the original diamond net are related by a $\bar{4}$ instead of a simple 4-fold axis of symmetry. The three bonds meeting at each point may now be (equal and) coplanar, and the geometry is defined by the following relationships, where ϕ is the angle between adjacent sides of the buckled 4-gon (originally 90°):

$$z = \frac{x}{4 - 8x} \qquad c/a = \sqrt{\frac{(1 - 8x + 8x^2)(4 - 8x)}{6x - 1}}$$

$$\tan \alpha = \frac{c/a}{2 - 4x} = \frac{2z}{x} \cdot \frac{c}{a} \qquad \sin \frac{\phi}{2} = \frac{2x \cos \alpha}{1 - 4x}$$

The upper limit of x (and of z, c/a, and α) is reached when the four points around each node of the diamond net form a regular tetrahedron (case c below); the net is now 4-connected. At the other extreme the whole net would become compressed into the plane 4.8^2 net (case a), but before this stage is reached, the distance between points related by a translation of c would be equal to the length of a link of the net. The condition for this to occur ($\sin \alpha = \frac{1}{4} - 2z$) sets the lower limit of α at around 9°. Figure 6.11 shows partial

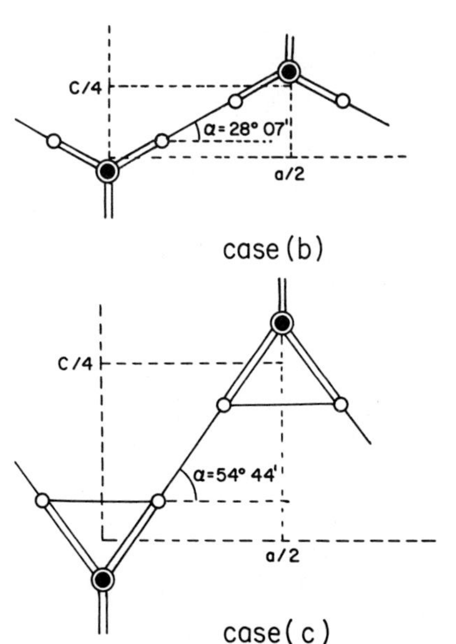

Fig. 6.11. Elevations of two configurations of the net 4.14^2.

elevations [on (100)] of two configurations of this net, namely, the regular tetrahedral model case *c* and an intermediate model case *b*; case *a* (not shown) corresponds to the plane net:

	Case *a*	Case *b*	Case *c*
x	0.146	0.150	0.159
z	0	0.054	0.058
c/a	0	0.748	1.93
ϕ	90°	76°	60°
α	0°	28°	54°44′

The configuration of Fig. 6.9 of the net 4.14^2 arises by placing squares (with their planes parallel) at the nodes of the simplest 3D 4-connected net. The analogous configuration (Fig. 6.12*a*) of the net 6.10^2 arises by placing hexagons (with their planes parallel) at the nodes of the simplest 3D 6-connected net, that is, the primitive lattice. This configuration of the net is also illustrated as a stereo-pair in Fig. 6.13. A configuration of the net with nonplanar hexagonal rings was mentioned in the derivation of the uniform 3-connected net (12, 3) on page 36. An alternative mode of linking pairs of intertwining 6_1 helices set up at the corners of a hexagonal cell leads to a system of two identical interpenetrating nets 6.10^2.

As noted earlier, the geometry of this net has certain similarities to that of the net 4.14^2. In the model with planar hexagons the angle of slope of the "slanting" links to the horizontal may be varied and/or the hexagons may be buckled. Assuming that unconnected points of the net should not be closer together than those in the net, the upper limit for buckled hexagons corresponds to the formation of octahedra and a 5- rather than a 3-connected net. As in the case of 4.14^2, there is a range of configurations in which all bonds have equal length and are coplanar at each point. Using the same symbols as for the tetragonal net and referring to hexagonal axes, the geometry of these forms with equal coplanar bonds can be summarized as follows:

$$z = \frac{x}{4 - 18x} \qquad c/a = \sqrt{\frac{(27x^2 - 12x + 1)(2 - 9x)}{15x - 2}}$$

$$\tan\alpha = \frac{c}{a}\cdot\frac{4z}{x} = \frac{c}{a}\cdot\frac{2}{2 - 9x} \qquad \sin\frac{\phi}{2} = \left(\frac{\sqrt{3}}{2}\right)\cos\theta$$

where $\tan\theta = \frac{1}{2}\tan\alpha$.

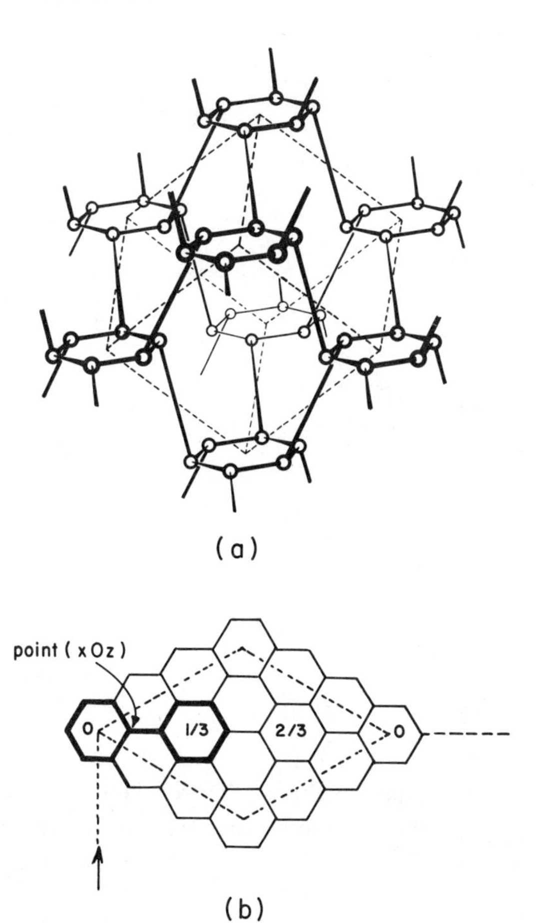

Fig. 6.12. (*a*) The net 6.10^2. (*b*) Projection on (0001). Arrow indicates direction of view for the elevations of Fig. 6.14 (portion drawn with heavy lines).

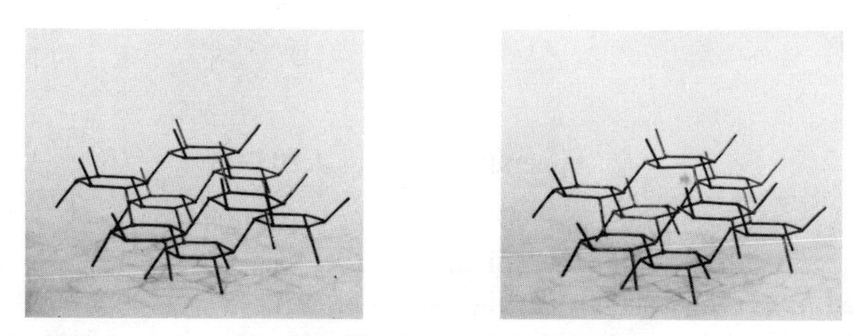
Fig. 6.13. The 3-connected net 6.10^2.

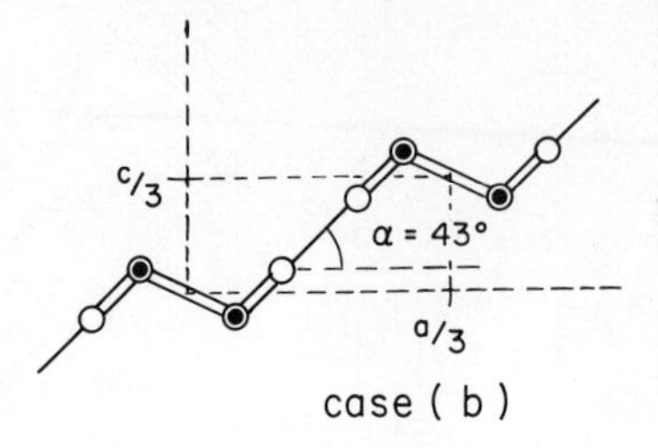

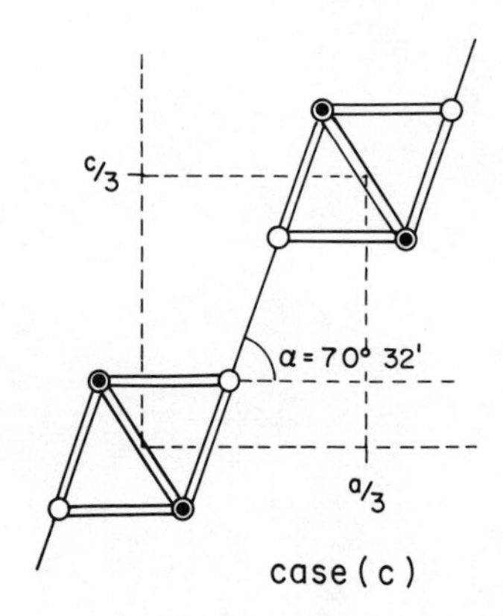

Fig. 6.14. Elevations of two configurations of the net 6.10^2.

Figure 6.14 shows partial elevations of the regular octahedral (5-connected) net, case *c*, and an intermediate model, case *b*; case *a* (not shown) corresponds to the plane 6-gon net:

	Case *a*	Case *b*	Case *c*
x	0.111	0.120	0.1276
z	0	0.065	0.075
c/a	0	0.43	1.20
ϕ	120°	103°	60°
α	0°	43°	70°32′

Appendix: the net 4.10^2

We have noted that the 3-connected systems $3.n^2$ are known for $n = 3$ and for all even values from 4 to 24 inclusive except 22. The net 3.22^2 would be derived by placing triangles at the points of a 3D 3-connected net 11^3, or (11, 3), and such a net is not known. The gap in this series occurs between two

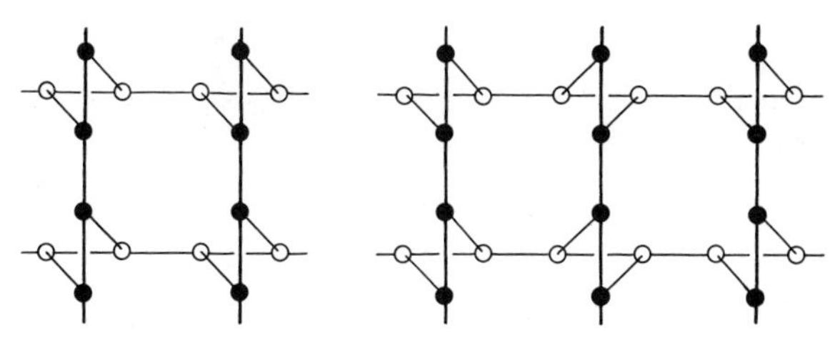

Fig. 6.15. Forms of the "double layer" net 4.10^2: points shown as filled and open circles lie in two parallel planes.

3D nets, but if we list the systems $4.n^2$ (n even) that have been illustrated, it is noticed that 4.10^2 has not been described. This net occurs between the 2D net 4.8^2 and the 3D net 4.12^2:

4.4^2	4.6^2	4.8^2	(4.10^2)	4.12^2	4.14^2
Finite		2D		3D nets	

The net 4.10^2 can be realized as a "double layer," which is illustrated, with a variant, in Fig. 6.15. Attempts to find additional members of this family with higher values of n have not been successful, although nets 4.12.18, 4.14.18, and 4.16.18 can be derived from the "quartz net" $8^2.6^4$ (p. 131)—compare the Archimedean solids 4.6.8 and 4.6.10.

7

Edge-sharing octahedral AX_3 structures derived from 3-connected nets

The well-known layer structures of composition AX_3 built from octahedral AX_6 groups sharing a symmetrical selection of three edges (YCl_3 and BiI_3 structures) are based on the simplest 2D 3-connected net (Fig. 7.1*a*). In these crystals the packing of the X atoms is respectively c.c.p. or h.c.p., but more complex layer sequences are possible. There are also octahedral edge-sharing AX_3 structures derivable from certain uniform 3D 3-connected nets, which can be constructed with equal coplanar links inclined at 120°. This is a necessary but not sufficient condition for forming such a structure. For

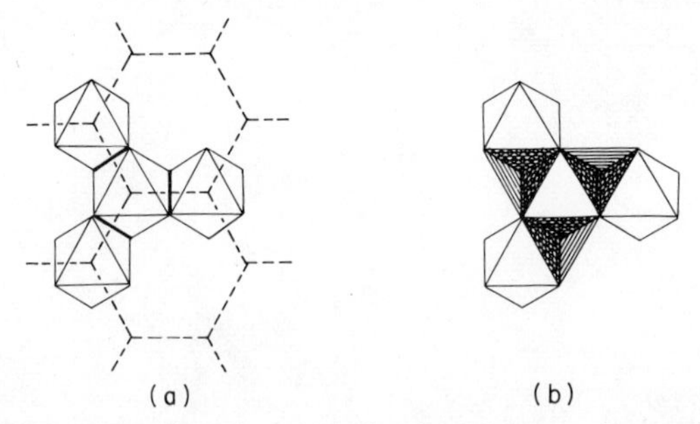

Fig. 7.1. (*a*) Formation of AX_3 layer based on the plane 6-gon net by octahedra AX_6 sharing three edges. (*b*) Addition of tetrahedra (shaded faces) to layer *a*—see text.

example, such an AX_3 structure can be built from the net (8, 3)-*b* but not from (8, 3)-*a*, or from (10, 3)-*c*, because of incompatible angular relations between edges of octahedra that would have to be shared. (This point can be appreciated easily only from a model.) The three known structures of this kind are as follows:

3D net	Packing of X atoms	Figure
(i) (10, 3)-*a*	c.c.p. ($\frac{1}{4}$ missing)	7.3
(ii) (10, 3)-*b*	c.c.p.	7.4
(iii) (8, 3)-*b*	c.c.p. ($\frac{7}{25}$ missing)	7.5

The positions of the missing X atoms in one layer of each of the structures (i) and (iii) are shown in Fig. 7.2. In (i) 4 out of 16 X positions in each layer are unoccupied, and in the structure as a whole these missing X atoms would delineate one of the four (10, 3)-*a* nets into which cubic closest packing can be broken down. In (ii) 7 of the 25 positions in each layer are unoccupied, leaving 18 for every 6 A atoms.

With the structures (i)–(iii) we may compare the 3D octahedral vertex-sharing AX_3 structure based on the simplest 3D 6-connected net. In the most symmetrical form of this structure (the cubic ReO_3 structure) the X atoms occupy, as in structure (i), three-quarters of the positions of cubic closest packing, but the octahedra may be rotated relative to one another so that the X atoms occupy *all* the positions of hexagonal closest packing; intermediate

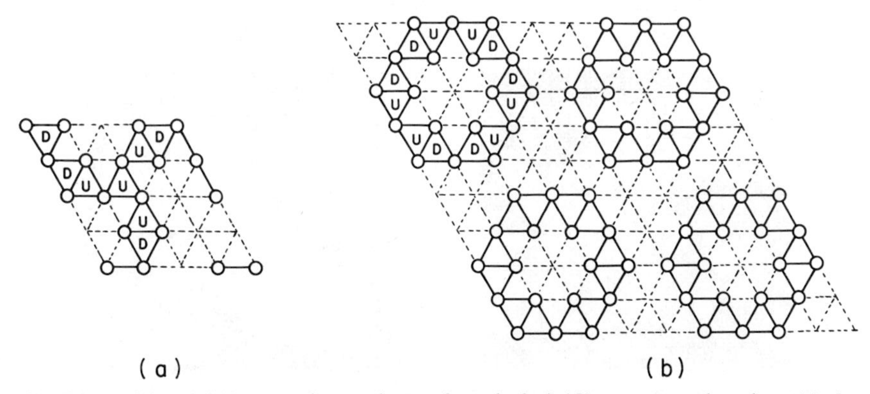

Fig. 7.2. Positions of X atoms in one layer of octahedral AX_3 structures based on (*a*) the net (10, 3)-*a* and (*b*) the net (8, 3)-*b*. Each triangle represents a face of an octahedron the remaining vertices of which are in a layer above (*U*) or below (*D*) that of the diagrams.

Fig. 7.3. Octahedral AX_3 structure based on the net (10, 3)-*a*. The spheres indicate the positions of the "missing" c. p. atoms.

Fig. 7.4. Octahedral AX_3 structure based on the net (10, 3)-*b*.

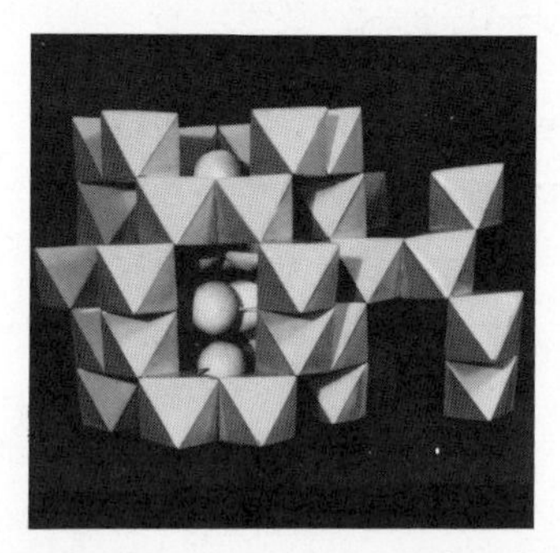 

Fig. 7.5. Octahedral AX_3 structure based on the net (8, 3)-*b*. The spheres indicate the positions of the "missing" c. p. atoms.

configurations are also possible. In contrast to the numerous compounds crystallizing with the vertex-sharing (ReO_3–RhF_3) type of structure, no examples appear to be known of compounds with the 3D edge-sharing structures.

The edge-sharing octahedral structures (i)–(iii) are related to other 3D nets and to 3D polyhedra. Tetrahedra may be added to the AX_3 structure (i) as indicated by the shaded areas in Fig. 7.1*b*. Each tetrahedron shares two faces (more heavily shaded) with octahedra of the framework, with the result that the tetrahedra and octahedra together form a 3D polyhedron [$(O_{3t}$-(3, 8)] which is described in Chapter 17. The tetrahedra themselves form a continuous vertex-sharing structure in which each vertex is common to two tetrahedra. The composition is therefore BX_2 if the structure is built from BX_4 tetrahedra. The B atoms are arranged at the points of the 4-connected $3^2 10^4$ net (p. 127), and this tetrahedral BX_2 structure is a special configuration having the X atoms in three-quarters of the positions of cubic closest packing. The octahedral AX_3 structure of Fig. 7.3 is unique in giving rise in this way to a regular 3D polyhedron and also to a tetrahedral BX_2 structure of the silica type based on a 4-connected net. Addition of tetrahedra to the AX_3 structures of Figs. 7.4 and 7.5 does indeed lead to 3D polyhedra and to 3D BX_2 structures, but these are less simple. In the tetrahedral structure derived from (ii) the tetrahedra are of two kinds, corresponding to the nonequivalence of the two kinds of link in the net (10, 3)-*b*. Two-thirds of the tetrahedra are of type *a* and one-third of type *b*, where the numbers are the

(*a*) $BX_{13/6}$ (*b*) $BX_{5/3}$

numbers of tetrahedra to which each vertex is common. The composition of the 3D structure is BX_2, since $(BX_{13/6})_2 + BX_{5/3} = 3(BX_2)$. Similarly the vertices of the corresponding 3D polyhedron are not equivalent (Fig. 7.6), having seven, eight, or nine triangles meeting at the three kinds (c_7, c_8, and c_9) of nonequivalent vertex. This type of 3D polyhedron, like others with mixed vertex types, c_5 and c_8 (p. 207), c_6 and c_7 (p. 214), and c_6 and c_8 (p. 219), is analogous to the eight *finite* convex polyhedra that have all their faces *equilateral* triangles but have different numbers of faces meeting at various vertices (Table 7.1). These are some of the plane-faced convex polyhedra that

Fig. 7.6. 3D polyhedron formed by addition of tetrahedral tunnels to the AX_3 structure of Fig. 7.4.

Table 7.1 The finite convex polyhedra with equilateral triangular faces*

Polyhedron	Total number of faces	c_3	c_4	c_5
Tetrahedron	4	**4**	—	—
Trigonal bipyramid	6	2	3	—
Octahedron	8	—	**6**	—
Pentagonal bipyramid	10	—	5	2
Bisdisphenoid	12	—	4	4
Tricapped trigonal prism	14	—	3	6
Bicapped square antiprism	16	—	2	8
Icosahedron	20			**12**

* There is no triangulated 18-hedron with equilateral faces. Regular (Platonic) solids are in boldface type.

can be constructed with equilateral regular faces; they comprise 110 polyhedra plus the two infinite families of prisms and antiprisms.* Again we see the close analogy between finite and 3D polyhedra—compare the "Platonic" 3D polyhedra (Chapter 17) with all vertices of the same type with polyhedra such as that of Fig. 7.6, with vertices of the types c_7, c_8, and c_9.

* See, for example, M. Berman, *J. Franklin Inst.*, 1971, **291**, 329.

8

Three-dimensional (3,4)-connected nets

For 3D nets the simplest (topological) repeat unit is the linear system of three points intermediate between the units for 3-connected and 4-connected nets (Fig. 4.1*a*). In diagrams of 3D (3, 4)-connected nets the 4-connected points are conveniently distinguished as shaded circles. The minimum value of Z is here the same as for a 2D net, in contrast to 3- and 4-connected nets, for which the minimum values of Z for plane and 3D nets are, respectively, 2 and 4, and 1 and 2. No attempt has been made to derive the 3D (3, 4)-connected nets systematically. Since the number of 3-connected points (c_3) in a repeat unit must be even, there are presumably nets of the following types:

Z	c_3	c_4
3	2	1
4	2	2
5	4	1
—	2	3

Conceivably there may be more than one net corresponding to a particular combination of values of c_3 and c_4.

We describe first seven uniform nets of which four form a special family, since they result from replacing one or more of the (4-connected) points in the diamond net by a pair (pairs) of 3-connected points. A fifth net, $\left(8, \begin{matrix}3\\4\end{matrix}\right)$-*a* of Table 8.2, is of interest for a different reason. Special interest attaches to nets in which the 3-connected and 4-connected points alternate, since if the points represent X and A atoms, such nets are possible structures for compounds A_3X_4. In these nets the ratio c_3/c_4 is equal to 4/3, and Z must be a multiple of 7. No example of a net with $Z = 7$ has been found, but of the two known with $Z = 14$, one (Fig. 8.9) is uniform; the other, which represents the phenacite (Ge_3N_4) structure, is not uniform. We include the latter with other nonuniform nets.

Uniform (3, 4)-connected nets

All seven nets described here are uniform in the strict sense of the term; that is, the shortest circuit including any pair of links from any point is an n-gon. The symbol of the net is $(n^3)_a(n^6)_b$, where a and b indicate the relative numbers of 3- and 4-connected points. As regards the value of n, these nets overlap the uniform 3- and 4-connected nets:

	n					
3-connected	—	7	8	9	10	12
(3, 4)-connected	6	7	8	9	—	—
4-connected	6	—	—	—	—	—

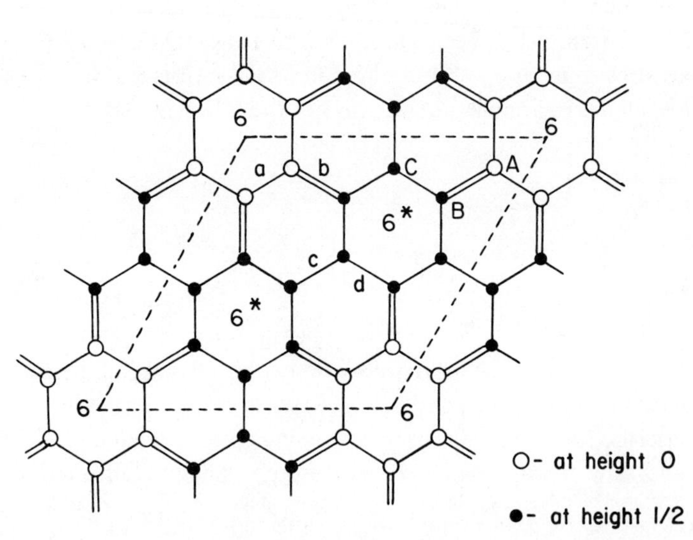

Fig. 8.1. The net $(6^3)(6^6)_2$ projected on (0001). The numerals 6 and 6^x indicate the two kinds of nonequivalent 6-gon (at heights 0 and $\frac{1}{2}$).

The only $(6^3)_a(6^6)_b$ net yet known was discovered in the course of studying 4-connected nets. It was observed that there is such a net which is closely related to the radiating (6^6) net of Fig. 9.13. Like the latter, it projects as the the plane 6-gon net (Fig. 8.1). This net is described as follows:

Hexagonal Space group $P6/mmm$ (No. 191) $Z = 18$

six 3-connected points in 6 (m), $(x_1, 2x_1, \frac{1}{2})$, C
six 4-connected points in 6 (m), $(x_2, 2x_2, \frac{1}{2})$, B } in Fig. 8.1
six 4-connected points in 6 (l), $(x_3, 2x_3, 0)$, A

The topological description is as follows. The ratio $c_3/c_4 = 1/2$. There are three kinds of nonequivalent points and four types of link:

Point	x	Link	y	Number per cell
A	9	$a\ (A–A)$	5	6
B	7	$b\ (A–B)$	4	12
C	5	$c\ (C–C)$	4	3
		$d\ (B–C)$	3	12
x_{mean}	7	y_{mean}	$\frac{42}{11}$	$p_{\text{mean}} = \frac{11}{3}$

The net is illustrated as a stereo-pair in Fig. 8.2.

Successive replacement of 4-connected points of the diamond net by pairs of 3-connected points gives a family of (3, 4)-connected nets and finally the (10, 3) nets described earlier. There are alternative ways of connecting the pairs of 3-connected points to the 4-connected ones, but only in certain cases are the resulting (3, 4)-connected nets uniform. It appears that the family consists of one uniform $\left(7, {3 \atop 4}\right)$, two uniform $\left(8, {3 \atop 4}\right)$, and one uniform $\left(9, {3 \atop 4}\right)$ (Table 8.1). These nets are all consistent with the relation

$$(n - 2)(p - 2) = 8$$

where p is the weighted mean connectedness. Apparently none of these nets can be constructed with the most symmetrical arrangements of links from each point, namely, three coplanar or four (regular) tetrahedral.

Figure 8.3 shows the relation of the four intermediate nets of Table 8.1, namely, $\left(7, {3 \atop 4}\right)$, $\left(8, {3 \atop 4}\right)$ (two), and $\left(9, {3 \atop 4}\right)$, to the b.c. tetragonal unit cell

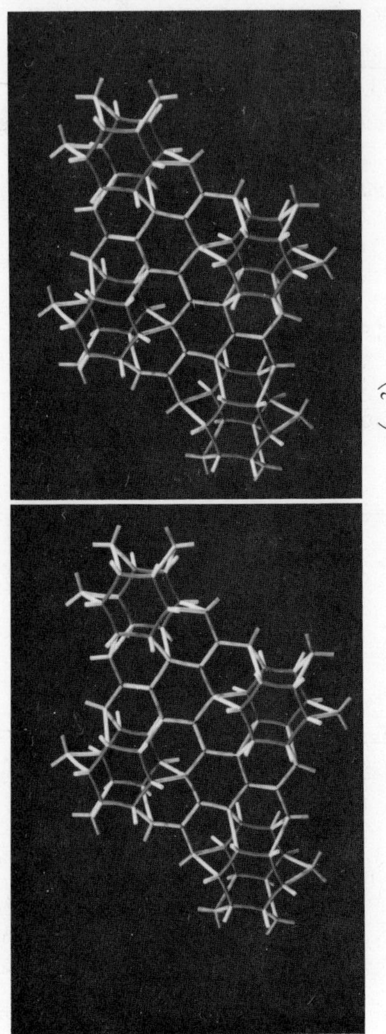

Fig. 8.2. The uniform net $\begin{pmatrix} & 3 \\ 6, & \\ & 4 \end{pmatrix}$.

Table 8.1 The diamond family of nets

Z	c_3	c_4	n	p_{mean}
4	0	4	6	(4)
5	2	3	7	$\frac{18}{5}$
6*	4	2	8	$\frac{10}{3}$
7	6	1	9	$\frac{22}{7}$
8	8	0	10	(3)

*Two different uniform nets $\begin{pmatrix} & 3 \\ 8, & \\ & 4 \end{pmatrix}$.

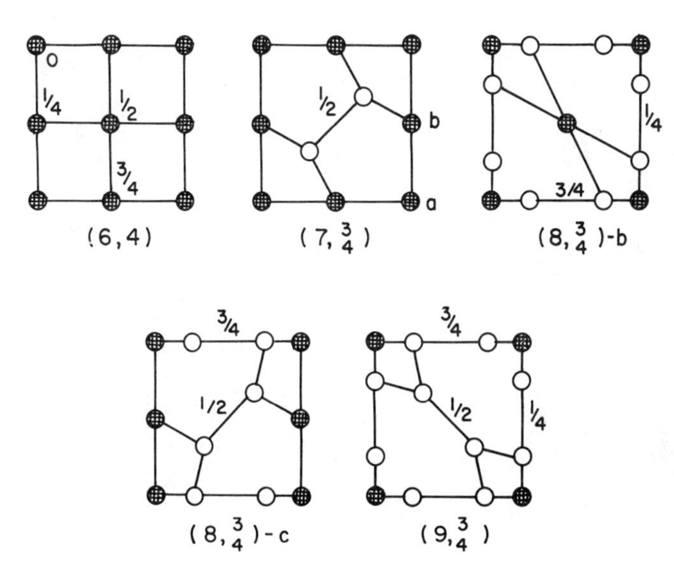

Fig. 8.3. The diamond net (b-c. tetragonal cell) and the derived (3, 4)-connected nets. The nomenclature is that of Table 8.1.

of the diamond net ($Z = 4$). The nets $\left(7, \frac{3}{4}\right)$ and $\left(9, \frac{3}{4}\right)$ were in the *Acta Crystallographica* series illustrated only as topological diagrams (Part 8, Fig. 7); they are presented as stereo-pairs in Figs. 8.4 and 8.5.

The difference between the two $\left(8, \frac{3}{4}\right)$ nets is evident from Fig. 8.3. One, $\left(8, \frac{3}{4}\right)$-*b*, contains isolated 4-connected points and isolated pairs of 3-connected points; it is illustrated in Fig. 8.6*a*, and a projection along the direction *XY* in that figure is shown in Fig. 8.6*b*. This projection is of interest in connection with the structure of the δ phase in the molybdenum-nickel system.[1] Figure 8.6*a* shows the most symmetrical configuration of the net $\left(8, \frac{3}{4}\right)$-*b*:

Tetragonal Space group $I\bar{4}m2$ (No. 119) $Z = 6$

4-connected points: 2(*d*) $(0\frac{1}{2}\frac{3}{4})$, etc.

3-connected points: 4(*e*) $(00z)$, etc.

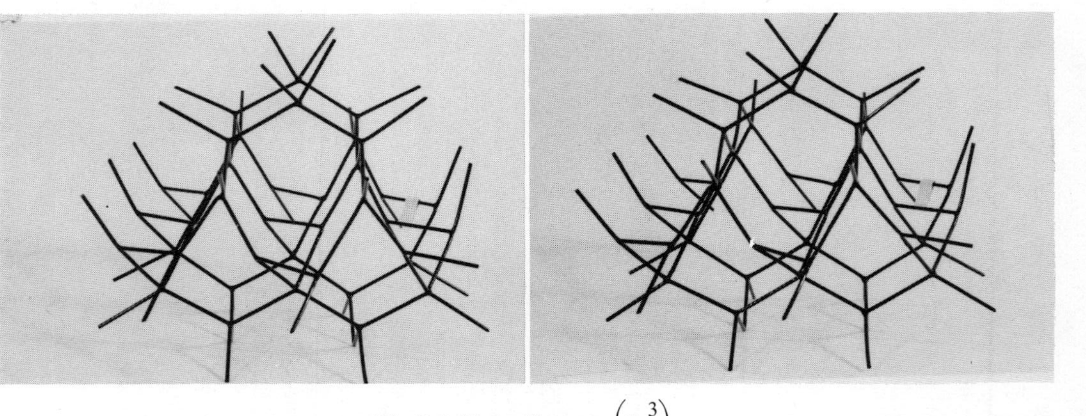

Fig. 8.4. The uniform net $\left(7, \genfrac{}{}{0pt}{}{3}{4}\right)$.

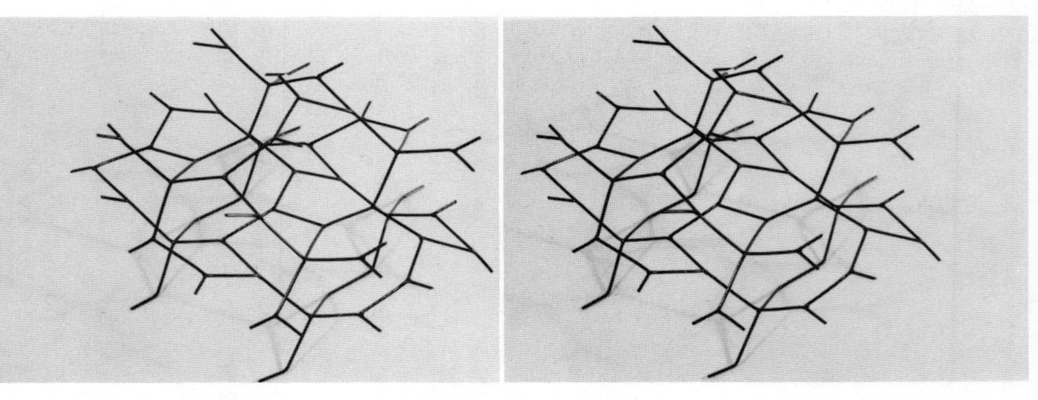

Fig. 8.5. The uniform net $\left(9, \genfrac{}{}{0pt}{}{3}{}\right.$

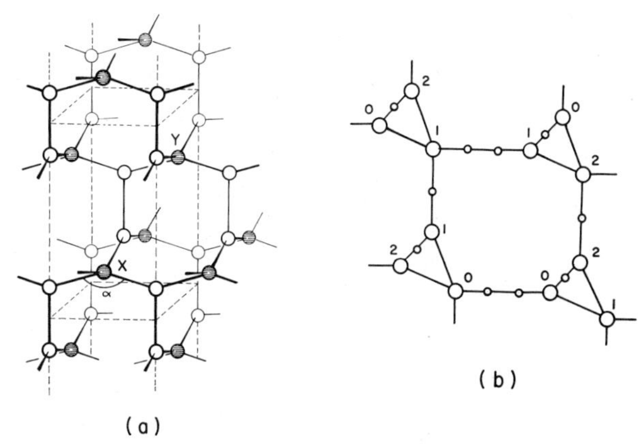

Fig. 8.6. The net $\left(8, {3 \atop 4}\right)$-*b*. (*a*) Unit cell; (*b*) projection along the direction *XY*. The smaller circles in (*b*) represent the 2-connected points in the "major skeleton" of the δ Mo-Ni phase.

It is not possible to have interbond angles of 120° for the 3-connected points *and* $109\frac{1}{2}°$ for the 4-connected points. Two possible configurations are

$$\alpha = 120°; \qquad z = \tfrac{1}{6}; \qquad c/a = \sqrt{3}$$

$$\alpha = 109\tfrac{1}{2}°; \qquad z = \frac{\sqrt{3}}{4\sqrt{3} + 8}; \qquad c/a = \frac{\sqrt{3} + 2}{\sqrt{2}}$$

The net $\left(8, {3 \atop 4}\right)$-*c* on the other hand contains continuous linked systems of 3-connected and of 4-connected points; these are more obvious in the alternative projection of Fig. 8.7 along one of the [100] axes of the "diamond" cell. The squares seen in the projection represent distorted 4-fold helices of 3-connected points; the directions of both these helices and the zigzag chains of 4-connected points, which project as pairs of shaded circles, are perpendicular to the plane of the paper. This net is illustrated as a stereo-pair in Fig. 8.8.

Topological data for the four (3, 4)-connected nets of Table 8.1 are given in Table 8.2, which includes the nets $\left(6, {3 \atop 4}\right)$, $\left(8, {3 \atop 4}\right)$-*a* and $\left(8, {3 \atop 4}\right)$-*d* that do not belong to the "diamond family."

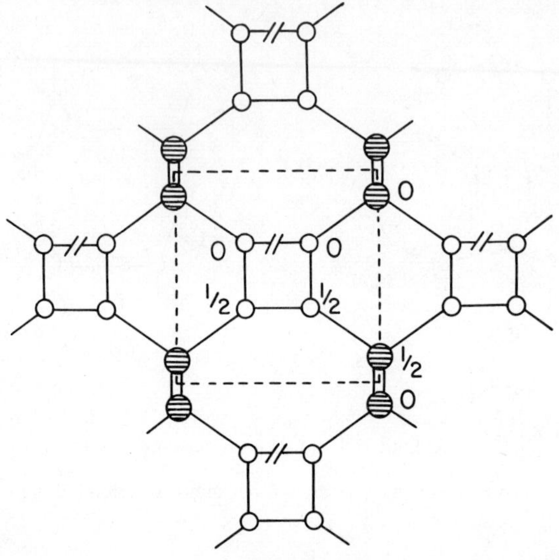

Fig. 8.7. Another projection of the net $\left(8, \begin{matrix}3\\4\end{matrix}\right)$-$c$, (cf. projection in Fig. 8.3).

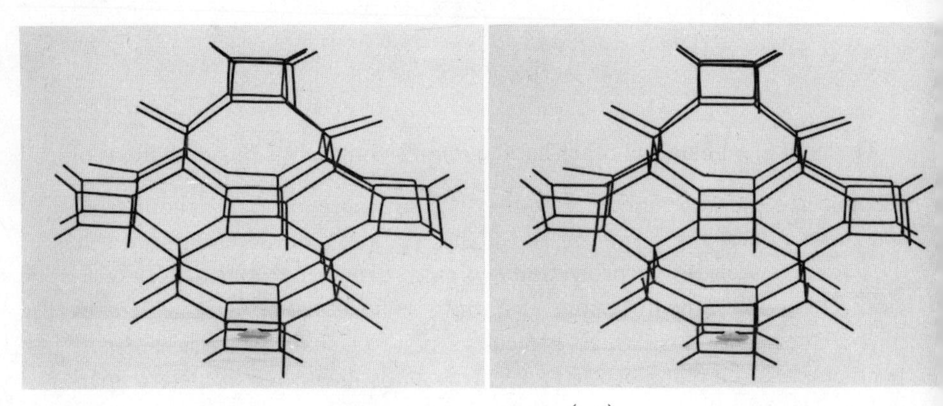

Fig. 8.8. The uniform net $\left(8, \begin{matrix}3\\4\end{matrix}\right)$-$c$.

Table 8.2 **Uniform (3, 4)-connected 3D nets**

Net	$x(c_3)$	$x(c_4)$	x_{mean}	y	y_{mean}	p_{mean}
$\left(6, \begin{matrix}3\\4\end{matrix}\right)$	5	9 (A) 7 (B)	7	A–A 5 (six) A–B 4 (twelve) C–C 4 (three) B–C 3 (twelve)	$\frac{42}{11}$	$\frac{11}{3}$
$\left(7, \begin{matrix}3\\4\end{matrix}\right)$	10 (two)	12 (a) 12 (b)	$\frac{56}{5}$	c_3–c_3 8 (one) c_3–c_4 6 (four) c_4–c_4 6 (four)	$\frac{56}{9}$	$\frac{18}{5}$
$\left(8, \begin{matrix}3\\4\end{matrix}\right)$-$a$ (Fig. 8.9)	15	20	$\frac{120}{7}$	10 (all links equivalent)	10	$\frac{24}{7}$
$\left(8, \begin{matrix}3\\4\end{matrix}\right)$-$b$	12	16	$\frac{40}{3}$	8 (two kinds of nonequivalent link)	8	$\frac{10}{3}$
$\left(8, \begin{matrix}3\\4\end{matrix}\right)$-$c$	12	16	$\frac{40}{3}$	8 (three kinds of nonequivalent link)	8	$\frac{10}{3}$
$\left(8, \begin{matrix}3\\4\end{matrix}\right)$-$d$ (Fig. 8.10)	9	12	$\frac{48}{5}$	6 (two kinds of nonequivalent link)	6	$\frac{16}{5}$
$\left(9, \begin{matrix}3\\4\end{matrix}\right)$	10 (two) 10 (four)	12 (one)	$\frac{72}{7}$	c_3–c_3 8 (three) c_3–c_3 6 (four) c_3–c_4 6 (four)	$\frac{72}{11}$	$\frac{22}{7}$

The net $\left(8, \begin{matrix}3\\4\end{matrix}\right)$*-a*

This net (Fig. 8.9) represents the arrangement of Pt and O atoms in $Na_xPt_3O_4^2$. The Pt atoms form four bonds to O and the O atoms three bonds to Pt.

Cubic Space group $Pm3n$ (No. 223) $Z = 14$

4-connected points: 6(c), $(\frac{1}{4}0\frac{1}{2})$, etc.
3-connected points: 8(e), $(\frac{1}{4}\frac{1}{4}\frac{1}{4})$, etc.

In this configuration of the net there is square planar bonding of the Pt atoms and equilateral planar bonding of the O atoms. This net is one of those mentioned earlier with $Z = 7m$, 3-connected and 4-connected points alternating. It is remarkable for the high x values of the points: that for the 3-connected points (15) is close to the highest value observed in 3-connected nets, and that for the 4-connected points (20) is higher than in any other net. Since y is the same for all links, these x values are necessarily in the ratio 3:4 because at any point $\sum y = 2x$.

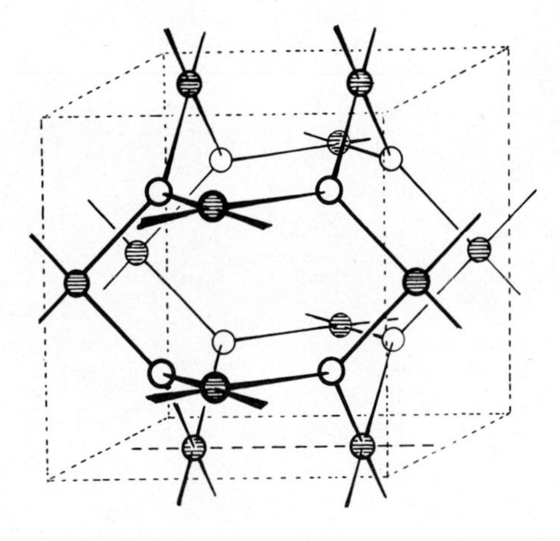

Fig. 8.9. The uniform net $\left(8, \begin{matrix}3\\4\end{matrix}\right)$-$a$, which represents the arrangement of 0 and Pt atoms in $Na_xPt_30_4$. Shaded circles represent Pt atoms. Na atoms at (000) and $(\frac{1}{2}\frac{1}{2}\frac{1}{2})$ are omitted.

The net $\left(8, \begin{matrix}3\\4\end{matrix}\right)$*-d*

This net (Fig. 8.10) has $Z_t = 5$ ($c_3 = 4$, $c_4 = 1$). A highly symmetrical configuration with equal links, interbond angles of 120° (c_3) and 90° (c_4), is

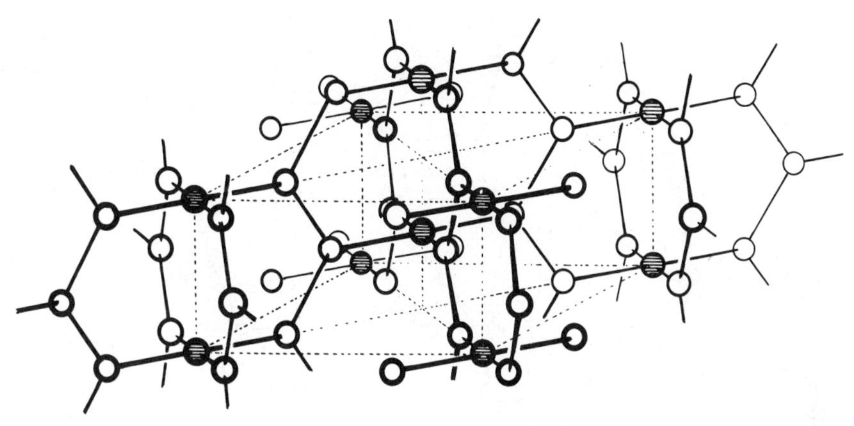

Fig. 8.10. The uniform net $\left(8, \begin{matrix}3\\4\end{matrix}\right)$-*d*.

fully described as follows:

Tetragonal Space group *I*4/*mmm* (No. 139) $Z = 10$ $c/a = \dfrac{\sqrt{6}}{5}$

4-connected points: 2(*a*), (000), etc.

3-connected points: 8(*h*), (*xx*0), etc.; $x = \frac{1}{5}$

For all the links $y = 6$. For the 3-connected points $x = 9$, and for the 4-connected points $x = 12$, giving $x_{\text{mean}} = \frac{48}{5}$. No example is yet known of a structure based on this net.

Other (3, 4)-connected nets

Two nonuniform (3, 4)-connected nets that represent crystal structures are illustrated in Figs. 8.11 and 8.12. The net of Fig. 8.11 represents the idealized structure of Ge_3N_4 and a number of complex oxides and fluorides A_2BX_4 in which A and B are tetrahedrally coordinated, including Be_2GeO_4, Be_2SiO_4 (phenacite), Zn_2SiO_4, Li_2MoO_4, Li_2WO_4, and Li_2BeF_4.

In $N(CH_3)_4F.4H_2O$, as reported by McLean and Jeffrey,[3] the F^- ions and H_2O molecules form a hydrogen-bonded 3D framework of the type appearing in Fig. 8.12. The coordinates of the (3-connected) points in 4-fold helices can be adjusted to give either (*a*) planar or (*b*) tetrahedral

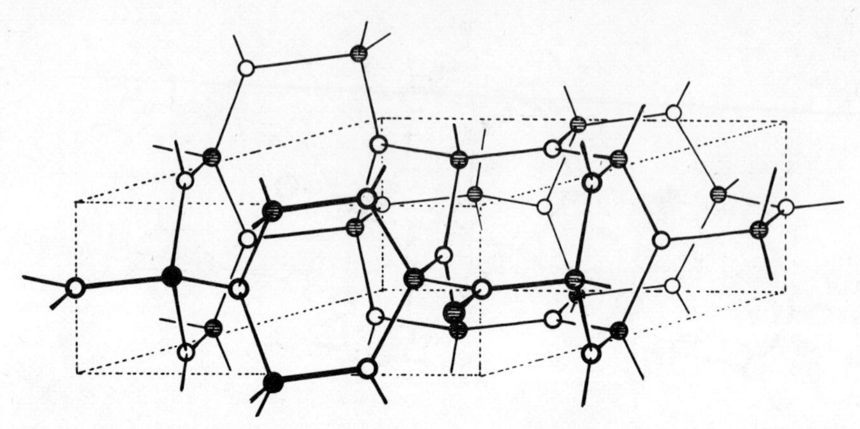

Fig. 8.11. The (3, 4)-connected net representing the structure of Ge_3N_4. Shaded circles represent Ge atoms in Ge_3N_4 (or Be or Si atoms in phenacite, Be_2SiO_4).

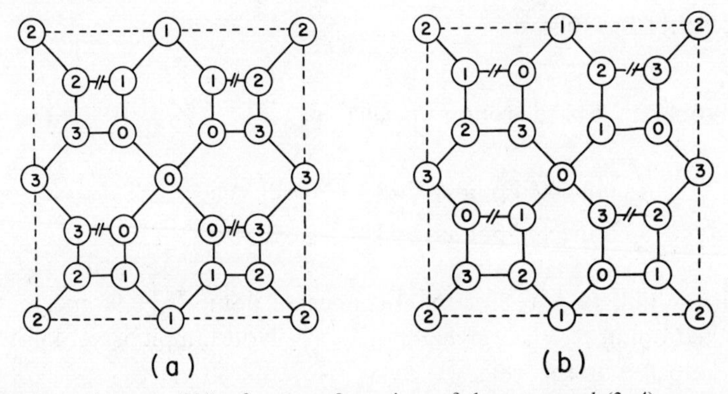

Fig. 8.12. Projections on (001) of two configurations of the tetragonal (3, 4)-connected net, an intermediate configuration of which represents the hydrogen-bonded framework of H_2O molecules and F^- ions in $N(CH_3)_4F.4H_2O$. Numerals indicate heights of points as multiples of $c/4$. The bond arrangement around 4-connected points is coplanar in (*a*) and tetrahedral in (*b*).

coordination around the 4-connected points. In $N(CH_3)_4F.4H_2O$ the coordination of the F^- ions is intermediate between the two extremes. (The cations are situated in the interstices of the framework.)

Three-dimensional borate ions

Networks built of 3- and 4-connected points are suitable for borates because by placing an O atom along each link, we may have planar BO_3 and tetra-

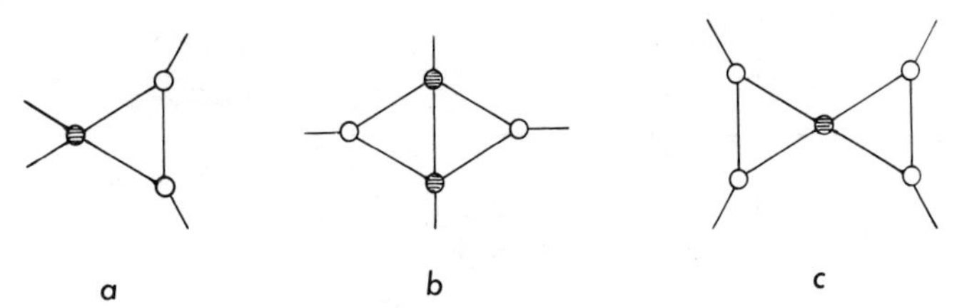

Fig. 8.13. Structural units in polyborate ions. (*a*) $(B_3O_5)_n^{n-}$; (*b*) $(B_4O_7)_n^{2n-}$; (*c*) $(B_5O_8)_n^{n-}$. Shaded circles represent 4-coordinated B atoms; O atoms are omitted.

hedral BO_4 groups joined by sharing O atoms. If L is the number of links in the repeat unit, $L = \frac{1}{2}(3c_3 + 4c_4)$, and the formula of such a 3D borate ion would be B_ZO_L (Z being the number of B atoms in the repeat unit); that is, B_3O_5, B_4O_7, B_5O_8, B_5O_9, B_5O_{11}, and so on. The charge on the ion is the value of c_4. Borates of these types are well known, and a number of structures have been determined. They do not, however, correspond to any of the nets described here. We have been interested in finding the nets with the largest circuits; the borates are built from subunits containing the smallest, most compact, ring systems. The O atoms along the links have been omitted from Fig. 8.13. Each unit has four free links and can form a net of the diamond type in which the repeat unit necessarily consists of two of the units of Fig. 8.13, so that $Z = 6$, 8, and 10, respectively. A further point of interest is that whereas in CsB_3O_5 the structure consists of one 3D framework ion (with the Cs^+ ions in the interstices) in $Li_2B_4O_7$, KB_5O_8, and other borates there are two interpenetrating frameworks as, for example, in Cu_2O.

References

1. C. C. Shoemaker and D. P. Shoemaker, *Acta Crystallogr.*, 1963, **16**, 997.
2. J. Waser and E. D. McClanahan, *J. Chem. Phys.*, 1952, **20**, 199.
3. W. J. McLean and G. A. Jeffrey, *J. Chem. Phys.*, 1967, **47**, 414.

9

Three-dimensional 4-connected nets

In 3D 3-connected nets the number (Z) of points in the repeat unit must be even, and the minimum value of Z is 4. In 3D 4-connected nets the minimum value of Z is 2, and all higher integral values are possible. We shall derive systematically the nets with 2, 3, and 4 points in the repeat unit, but we also refer to some more complex 4-connected nets that are of interest in connection with certain crystal structures.

When four or more links meet at each point in a net, there is the possibility that some or all of the links outline polyhedra:

1. There are discrete polyhedra.
2. All space is divided into polyhedra of one or more kinds.

If we proceeded with the derivation of nets to a sufficient degree of complexity we should include examples of the partitioning of space into polyhedral compartments. Reference is made to relevant aspects of this problem.

We reproduce here, with minor modifications, those portions of Part 2 describing the derivation of the first 20 3D 4-connected nets (those with $Z = 2, 3,$ or 4), but we wish to preface this section with a number of explanatory notes.

1. It is now known that the earlier work did not cover exhaustively even the limited field studied. The description of the method adopted could at least provide the basis for further studies.

2. It was implied (by the use of the word "permissible" in several places), though not explicitly stated, that if two points a distance a apart are connected, there should not be other pairs of points at the same distance apart that are *not* connected. The introduction of an apparently *geometrical* limitation into a supposedly *topological* derivation of nets has always appeared unsatisfactory, but it arises directly from the need for repetition by simple translation in a periodic 3D net.

3. In the derivation of the 4-connected 3D Nets 9–14 from (3, 4)-connected plane nets (section, From (3, 4)-connected nets) the plane nets with $Z = 4$ were those listed in Table 9.1. It is now known that there is another configuration of the net ($\phi_3 = \frac{2}{3}$, $\phi_8 = \frac{1}{3}$) and also of the net ($\phi_4 = \frac{1}{3}$, $\phi_5 = \frac{2}{3}$). These were not considered in the derivation of the 3D nets, and there may well be others. These two nets are included, as Figs. 14.6*e* and *f*. It is possible that they would lead to new 3D nets having $Z = 4$.

4. Reexamination of the derivation of the Nets 15–20 has already shown that the method does not necessarily lead to a unique net, suggesting that there may be other nets with $Z = 4$ that were not listed. Moreover, this complication also applies to Nets 2 and 3. It is to be expected that some of the possible ways of joining together the layers will only duplicate 3D nets already listed, but in any case the thorough examination of this problem would be time-consuming and would further delay the writing of this account of these studies. Reexamination

Table 9.1 Plane (3, 4)-connected nets and the derived 3D 4-connected nets

Z	ϕ_3	ϕ_4	ϕ_5	ϕ_6	ϕ_7	ϕ_8	3D 4-connected net (Fig. 9.2)
3	$\frac{1}{2}$	—	—	—	$\frac{1}{2}$	—	Net 2
	—	$\frac{1}{2}$	—	$\frac{1}{2}$	—	—	Net 3
4	$\frac{2}{3}$	—	—	—	—	$\frac{1}{3}$	Net 9 Net 10
	$\frac{1}{3}$	$\frac{1}{3}$	—	—	$\frac{1}{3}$	—	Net 11
	$\frac{1}{3}$	—	$\frac{1}{3}$	$\frac{1}{3}$	—	—	Net 12
	—	$\frac{2}{3}$	—	$\frac{1}{3}$	—	—	Net 13
	—	$\frac{1}{3}$	$\frac{2}{3}$	—	—	—	Net 14

of the derivation of Net 20 has given the results set out in a later section of this Chapter.

Systematic derivation of three-dimensional 4-connected nets with $Z = 2, 3,$ or 4

These may be derived by joining some or all of the points in plane nets to points in adjacent layers so that all the points become 4-connected. Plane nets of four types have to be considered.

1. From 3-connected nets: equal numbers of points in each layer must be joined to points in adjacent layers above and below.
2. From mixed 3- and 4-connected nets: the 3-connected points are to be joined either upward or downward to 3-connected points of adjacent layers. The number of 3-connected points in the repeat unit must clearly be even.
3. From mixed 2- and 4-connected nets: each of the 2-connected points must form two additional links to points in adjacent layers.
4. From mixed 2-, 3-, and 4-connected nets: the 2- and 3-connected points form additional links as in items 2 and 3.

The simpler 3- and 4-connected plane nets are listed in Tables 14.1 and 14.2 and illustrated in Figs. 14.3 and 14.5.

The 3D nets are listed in order of increasing numbers of points (Z) in the repeat unit (Table 9.2), but it is convenient to indicate their derivation under the headings just enumerated.

From 3-connected nets

The simplest plane 3-connected net is the hexagonal net with two points in the repeat unit. By connecting alternate points to points in layers above and below (Fig. 9.1*a*, *c*) there arises the simplest 3D 4-connected net (Net 1) with $Z = 2$. In Fig. 9.1 and similar figures open and filled circles represent points to be connected to points in layers above and below respectively; these are referred to as U and D points.

The numbers of points in repeat units of plane 3-connected nets are even. For $Z = 4$ there are the following possibilities:

1. Alternate layers A (Fig. 9.1*a*) and its "mirror image" A' (Fig. 9.1*b*) give the sequence of layers $AA'AA'\ldots$ shown in elevation in Fig. 9.1*d*

Table 9.2 3D 4-connected nets

Net	Z	Polygons in net 3	4	5	6	7	8	9
1	2	—	—	—	6	—	8	—
2	3	3	—	—	—	7	8	—
3	—	—	4	—	6	—	8	—
4	—	—	—	—	6	—	8	—
5	—	—	—	—	6	—	8	—
6	4	—	—	—	6	—	8	—
7	—	—	4	—	6	—	8	—
8	—	—	4	—	6	—	8	—
9	—	3	—	—	—	—	8	—
10	—	3	—	—	—	—	8	9
11	—	3	4	—	—	7	8	—
12	—	3	—	5	6	—	—	9
13	—	—	4	—	6	—	8	—
14	—	—	4	5	—	—	8	—
15	—	—	4	—	—	—	8	—
16	—	3	—	5	—	7	8	—
17	—	3	4	—	—	—	8	—
18	—	—	4	—	—	—	8	—
19	—	3	—	—	—	—	—	9
20	—	3	—	5	6	7	8	—
21	—	—	—	—	6	—	8	—

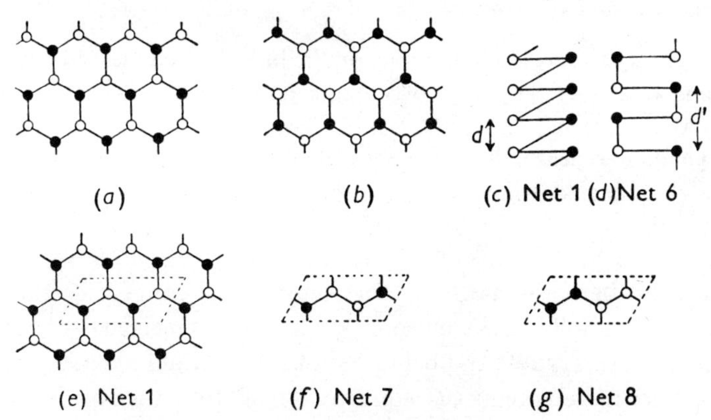

Fig. 9.1. Derivation of 4-connected nets (see text).

with four points in the repeat unit (Net 6). For tetrahedral bonds this is the hexagonal diamond (wurtzite) structure. There is an indefinite number of more complex sequences of A and A' layers; compare the SiC polytypes.

2. In a double repeat unit of the plane hexagonal layer we can arrange two D and two U points as in Figs. 9.1*e*, *f*, and *g*, where D represents a point to be connected downward to a U point of the layer below and U a point to be connected upwards to a D point of the layer above. Of these the first is simply Net 1; the others are new nets, Nets 7 and 8.

3. Plane 3-connected nets with four points in the repeat unit. No new nets arise from Nets 3 and 4 of Table 14.1.

For higher values of Z the 3D nets can be derived in a similar way. The $AA'AA'\ldots$ type of net occurs only for nets with $4n$ points in the repeat unit, because there are no plane 3-connected nets with odd numbers of points in the repeat units.

From mixed (3, 4)-connected nets

To form a 3D 4-connected net, there must be an even number of 3-connected points in the repeat unit. The simplest case is therefore the plane net with three points in the repeat unit, and the next simplest nets have $Z = 4$. The relevant plane nets are those of Table 9.1 and Figs. 9.2*a*, *b*, and *e*. (Figure 9.2*c* does not give rise to a permissible net because $x = y$.)

From mixed 2- and 4-connected nets

The plane nets arise by placing points on the links of 4-connected nets. There is no plane 4-connected net with two points in the repeat unit (see Table 14.2).

Two points in repeat unit. There is only one case: one 4-connected point and one 2-connected point, and this is not permissible because $x = y$ (Fig. 9.3*a*).

Three points in repeat unit. One net only: one 4-connected and two 2-connected points. In Fig. 9.3*b* the point A cannot be connected to corresponding points of layers above and below because this would make the distance between unconnected points CC' equal to AA' or BB', where primes indicate equivalent points in adjacent layers. There are, however, two other interesting

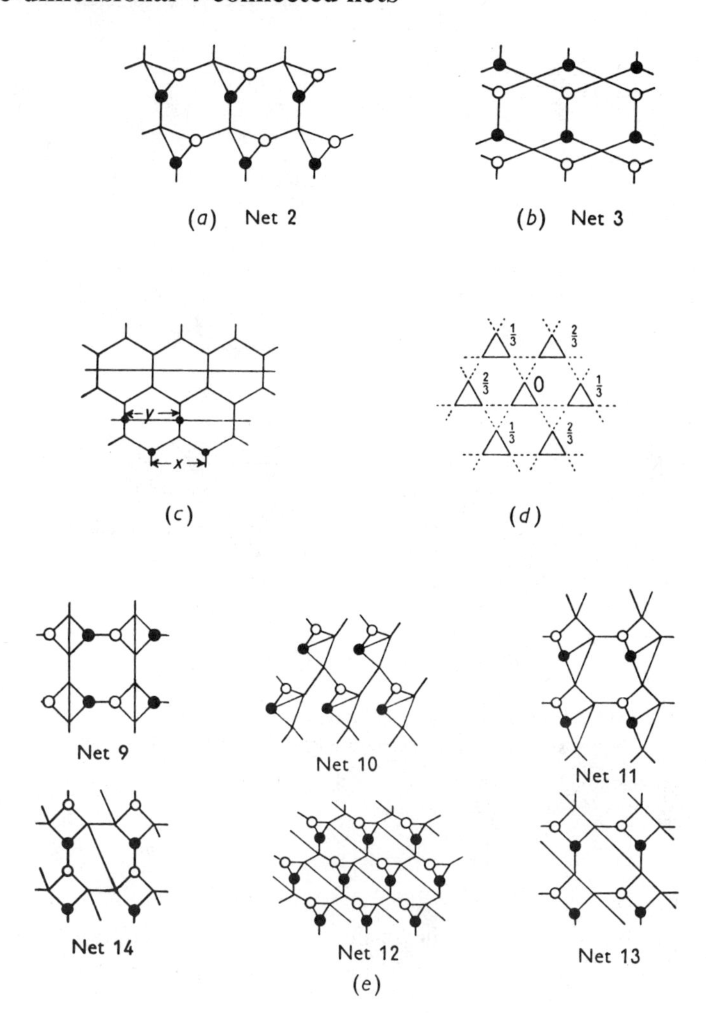

Fig. 9.2. (*a*) and (*b*) The 4-connected Nets 2 and 3. (*c*) See text. (*d*) Projection of rhombohedral form of Net 2 (heights as fractions of the length of rhombohedral [111] axis). (*e*) Derivation of the 4-connected Nets 9–14.

possibilities involving the linking of point *A* (Fig. 9.3*b*) of one layer to point *B*′ of the layer above and point *D*′ of either the layer below (Fig. 9.3*c*) or of the layer above (Fig. 9.3*d*). This mode of linking gives rise to screw axes (Fig. 9.3*e*, *f*), where the numbers indicate the heights of points above the plane of the paper in terms of $\frac{1}{3}$ (distance between layers). In Fig. 9.3*e* one-half the triangles represent clockwise helices and the other half anticlockwise

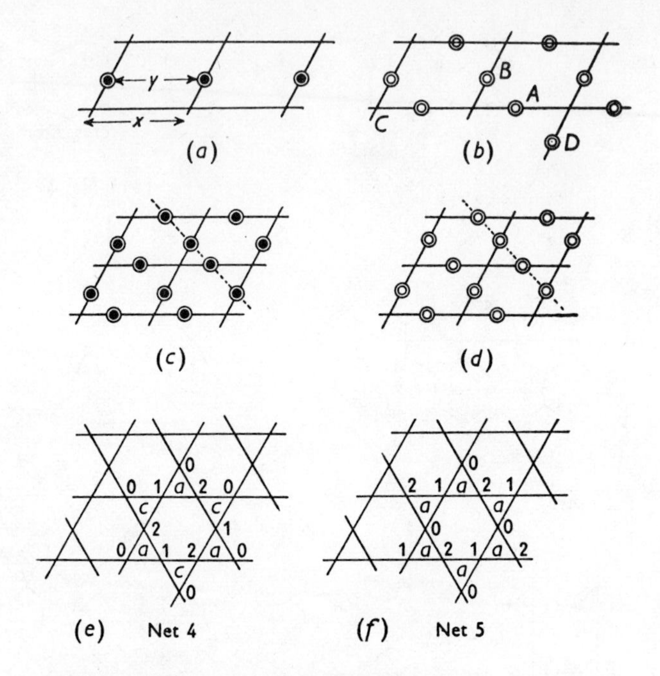

Fig. 9.3. Derivation of 4-connected nets (see text). Heights given as multiples of one-third the repeat distance between layers in (*e*) and (*f*).

helices. This net (Net 4*) can have rhombohedral or cubic symmetry. In Fig. 9.3*f* all the triangles in the projection represent helices of the same chirality (here anticlockwise); this is the enantiomorphic quartz net (Net 5).

Four points in repeat unit. No permissible nets arise.

From mixed 2-, 3-, and 4-connected nets

The relevant plane nets arise by placing (2-connected) points on the links of the plane (3, 4)-connected nets of Fig. 9.2. Since there must be an even number of 3-connected points in the repeat unit, the simplest nets are formed

* It is convenient to refer to Net 4 as the NbO net (this being the only compound with this simple structure), with the understanding that we mean the net with six equivalent points [6(*b*) $(0\frac{1}{2}\frac{1}{2})$ etc. in *Im*3*m* (No. 229)] and not two sets of nonequivalent points 3(*c*) and 3(*d*) in *Pm*3*m* as is necessarily the case for NbO. On this point see also page 131.

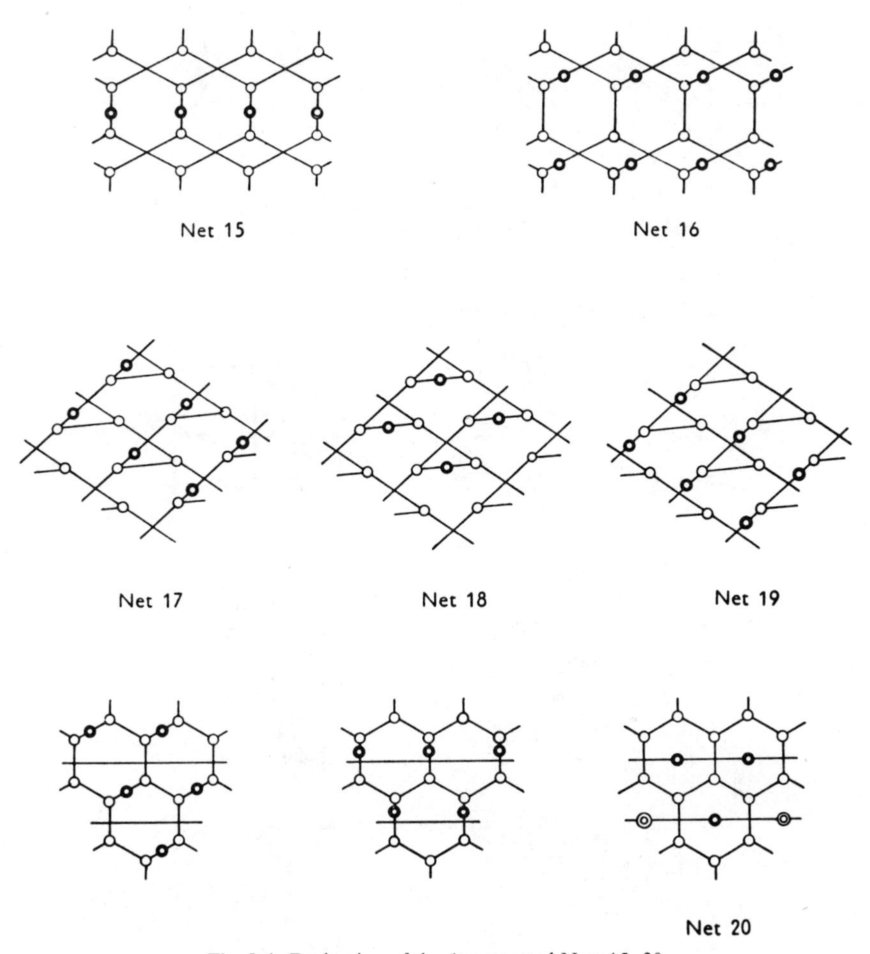

Fig. 9.4. Derivation of the 4-connected Nets 15–20.

from systems with four points in the repeat unit: one 2-connected, two 3-connected, and one 4-connected. The permissible nets are set out in Fig. 9.4. Of the three ways of placing the 2-connected point in the third net, only the third gives a permissible 3D net.

We have now derived the 4-connected nets with two, three, or four points in the repeat unit. They are summarized in Table 9.2, which lists the smallest polygons in the nets, and Nets 1–7 and 15 are illustrated, in their most symmetrical forms, in Fig. 9.5. Net 8 is illustrated as an open packing of tetrahedra in Fig. 9.32.

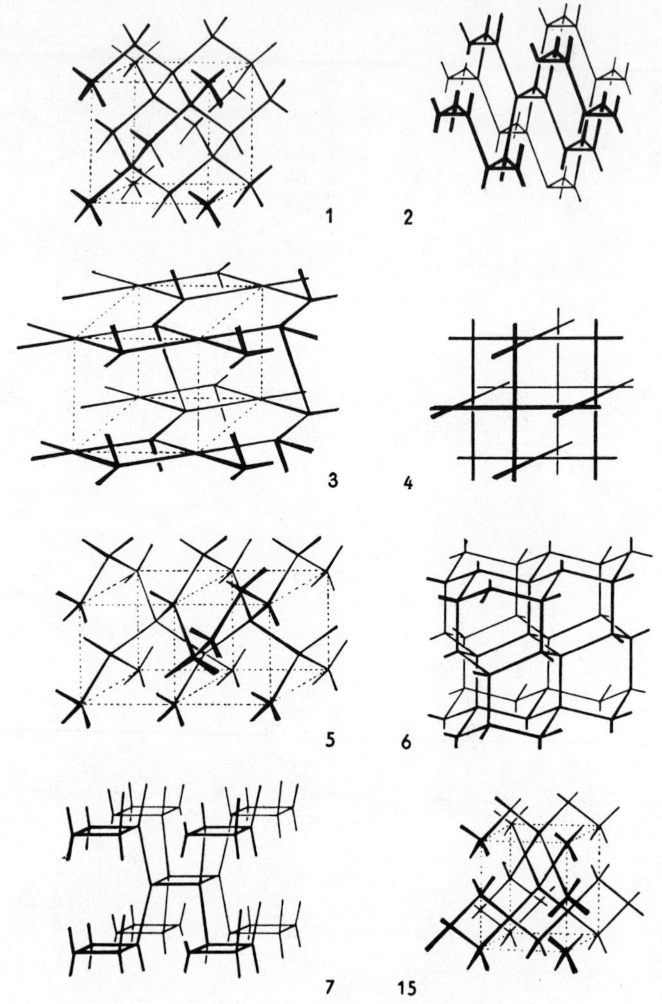

Fig. 9.5. Some 3D 4-connected nets. The numbering is that of Table 9.2.

Reexamination of "Net 20"

In the diagram labeled Net 20 in Fig. 9.4 the two connections from the 2-connected point could be made both up (or down) or one up and one down; moreover the connections could be made to different pairs of 3-connected points. Bearing in mind that $Z = 4$, the connections must be made as in Fig. 9.6*a* or *b*. In Fig. 9.6*a* we have to connect the 2-connected point (◎) to

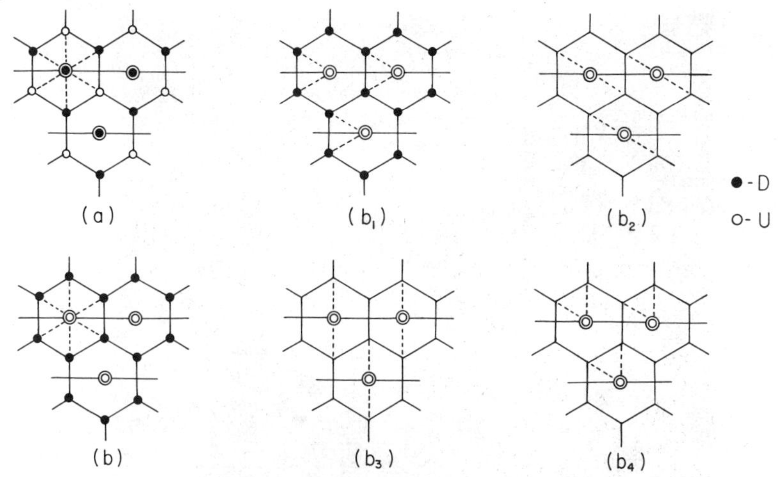

Fig. 9.6. Reexamination of "Net 20" (see text).

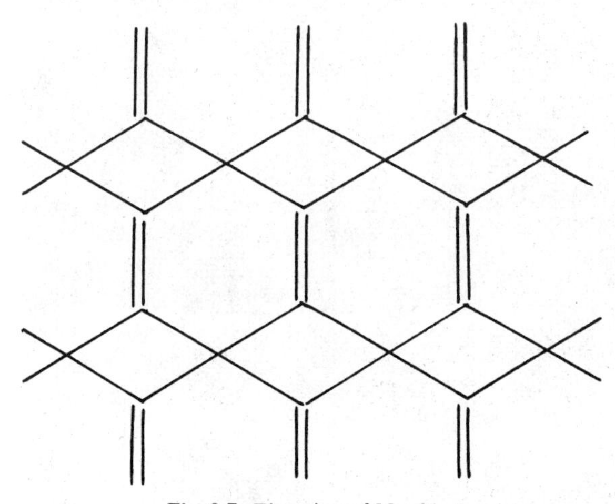

Fig. 9.7. Elevation of Net 21.

one D point (●) and one U point (○) of adjacent layers. The choice of any such pair of the six dotted lines in Fig. 9.6*a* leads to the same net, which we call Net 21, since it was not listed earlier. This new net, which has the elevation of Fig. 9.7, has equal numbers of points of the types 6^6 and $6^5 8$. The model shown as a stereo-pair in Fig. 9.8 is constructed with links of two lengths and with tetrahedral and coplanar arrangements of links around the non-equivalent points of types 6^6 and $6^5 8$. On the other hand Fig. 9.6*b* leads to

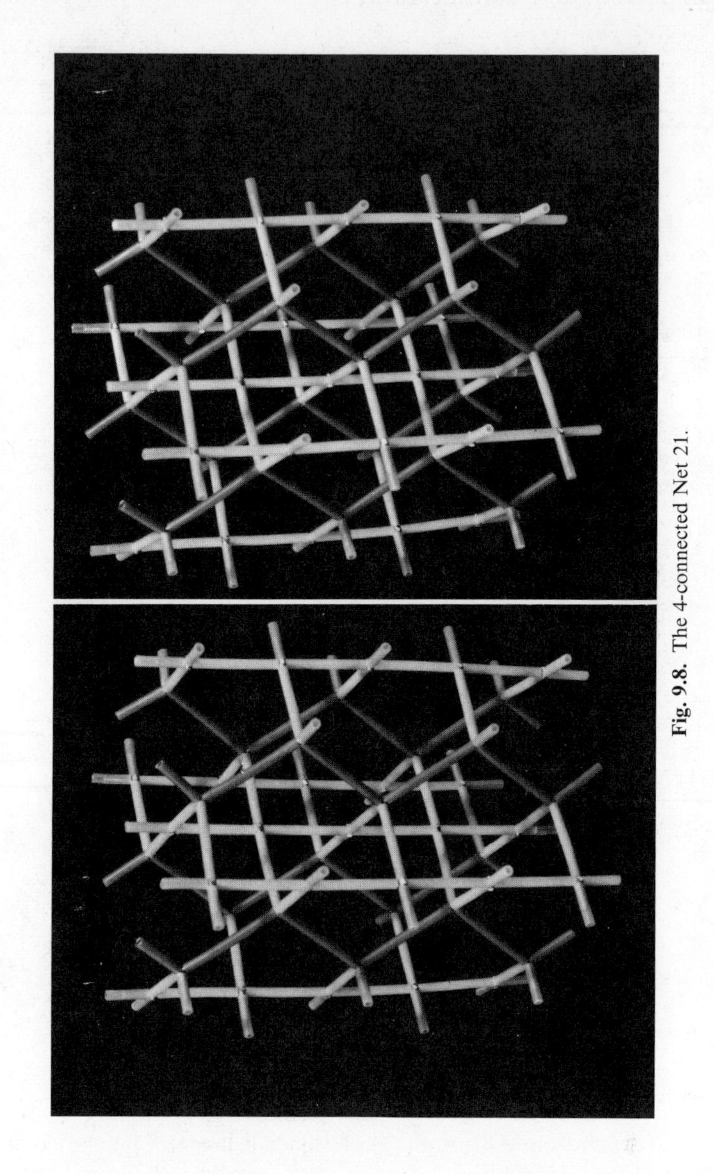

Fig. 9.8. The 4-connected Net 21.

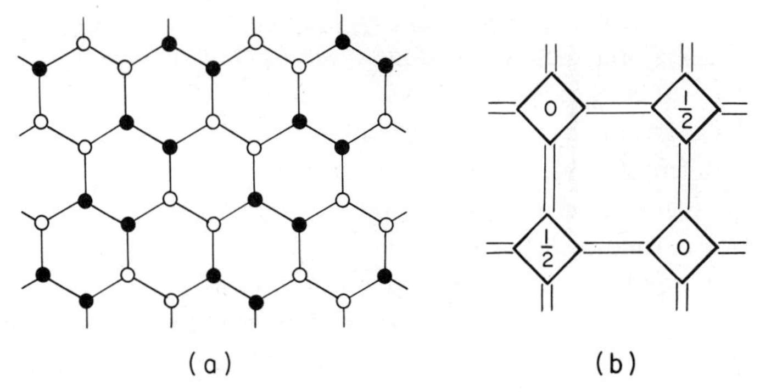

Fig. 9.9. Two projections of Net 7.

four ways of connecting the 2- and 3-connected points, and all correspond to different 3D nets (b_1–b_4):

Fig. 9.6b_1	Net 7
Fig. 9.6b_2	see below
Fig. 9.6b_3	Net 21
Fig. 9.6b_4	Net 20

Net 7 was originally derived as shown in Fig. 9.1*f*, repeated as Fig. 9.9*a*, and projects in one direction as Fig. 9.9*b*; this illustrates the difficulty of recognizing new nets and of being sure that they are not simply nets that have been previously derived in a different way. The net arising from b_2 appeared to be a new 6^6 net topologically different from those described earlier. However it is virtually certain that this net could not be constructed without bringing unconnected points closer together than connected ones, though this point has not been rigorously proved. Provisionally it is not be regarded as a "permissible" net.

The linking corresponding to b_4 is that originally intended to produce Net 20. This is a very complex net, for there are three kinds of topologically nonequivalent point with point symbols $3.5^2 6^3$ (one), $5.6^4 8$ (one), and $3.5.6^3 7$ (two). The original listing of circuits in this net was very incomplete; it is corrected in Table 9.2.

Further studies of 4-connected nets

Returning to this problem after a number of years, it appeared unprofitable to proceed further with the systematic derivation of nets with increasing numbers of points in the repeat unit. First, the nets become very numerous as Z increases, and their derivation would be not only tedious but not necessarily exhaustive, as noted earlier. Moreover, even some nets with small values of Z are complex in that they contain a variety of circuits of different sizes, and different combinations of circuits meet at nonequivalent points. It seemed preferable to concentrate first on the simplest nets, then to proceed to nets of greater topological complexity.

Nets with point symbol n^6 for all points are the 4-connected homologues of the *uniform* 3-connected nets n^3. A further reason for not continuing the systematic derivation of nets according to the value of Z is the finding that some topologically simple 3-connected nets have high values of Z, as shown for the 10^3 nets in Table 5.1. Unfortunately there seems to be no systematic way of finding such nets. For a long time it appeared that the only uniform 3D 4-connected nets were Nets 1 and 6 of Table 9.2, which represent the cubic and hexagonal diamond structures. It was observed recently that the structure of a cubic high-pressure form of silicon is also a 6^6 net that is related in an interesting way to Nets 1 and 6. Although we must admit the possibility that there exist other 6^6 nets, it still appears likely that the only uniform 4-connected nets are of the type 6^6.

In the following notes the nets are grouped according to their point symbols. The groups include some of the nets of Table 9.2 and also a number of more complex nets, more recently derived. Table 9.3 summarizes the

Table 9.3

3.6^5	4.6^5*	6^6-(*a*)
	$4^2.6^4$-(*a*)	6^6-(*b*)
	$4^2.6^4$-(*b*)	6^6-(*c*)
3.7^5		
		$6^4.8^2$-(*a*)
		$6^4.8^2$-(*b*)
$3^2.10^4$	$4^2.8^4$	
$3^3.12^3$	$4^3.6^3$	
	$4^3.8^3$	

* Two further 4.6^5 nets (Z = 8 and 24) have now been found.

periodic 4-connected nets of the types n^6 and $m^a n^{6-a}$ that have been discovered thus far, but because of the nature of the subject, it is unlikely that this list is complete.

Uniform nets n^6 and some related nets

Net 6^6-(a): the cubic diamond net

The only uniform 4-connected nets are n^6 nets, and of these the simplest is Net 1 of Table 9.2, which is also the simplest 3D 4-connected net and the only one with $Z_t = 2$. If this net is constructed with equal links and regular tetrahedral bond angles, the symmetry is cubic, space group $Fd3m$ (No. 227), $Z = 8$ in 8(*a*); it represents the structure of (cubic) diamond and isostructural crystals. This net may alternatively be referred to a b.c. tetragonal cell with $Z = 4$ or to a rhombohedral cell with $Z = 2$. All the shortest circuits in the most symmetrical form of this net are chair-shaped hexagons.

Extension or compression of the most symmetrical form of the diamond net leads to configurations that are of interest in relation to other topics. An extended configuration is mentioned in connection with interpenetrating nets (p. 150) and a compressed configuration referrable to a b.c. *cubic* unit cell is the basis of the 3D polyhedron (5, 5) that is also a uniform 5-connected net with the point symbol 5^9 (or $5^9 7$) (see p. 227). The alternative orthogonal cells for four special configurations of the diamond net are listed in Table 9.4.

Table 9.4 Special configurations of the diamond net

Net	Body-centered cell ($Z = 4$), c/a	Face-centered cell ($Z = 8$), c/a	Coordination type
Extended diamond	2.00	1.414	Elongated tetrahedral (4 + 4)
Cubic diamond	1.414	1.00*	Regular tetrahedral
Compressed diamond	1.00*	0.71	Flattened tetrahedral (4 + 2)
White tin†	0.515	0.374	Distorted octahedral (6 equidistant)

* Cubic; all other cells are tetragonal.

† Description refers to extreme configuration with six equidistant neighbors; for white tin, $c/a = 0.546$, corresponding to four neighbors at 3.02 Å and two at 3.17 Å.

Net 6⁶-(b)

Closely related to the cubic diamond net is Net 6 of Table 9.2, which in its most symmetrical configuration represents the structure of hexagonal diamond:

$$\text{Hexagonal} \qquad \text{Space group } P6_3/mmc\ (D_{6h}^4)\ (\text{No. } 194),$$

$$c/a = 1.633 \quad Z = 4 \text{ in } 4(f) \qquad z = \tfrac{1}{16}$$

The shortest circuits are of two kinds, chair- and boat-shaped hexagons. By rotating certain "layers" of points in this net about the c axis and remaking the appropriate links, an indefinitely large number of nets can be produced that are intermediate between the nets 6^6-(a) and (b); many are known as the structures of polytypes of SiC and ZnS.

The nets 6^6-(a) and (b) represent the positions of the centers of the spheres in the cubic and hexagonal variants of the sphere-packing H and $L\ 4_1$. The x and y values for these two nets are given later.

Net 6⁶-(c)

The structure of a more dense high-pressure form of silicon,[1] shown in projection in Fig. 9.10 and as a stereo-pair in Fig. 9.11, is described as follows:

$$\text{Cubic} \qquad \text{Space group } Ia3\ (T_h^7)\ (\text{No. } 206)$$

$$Z = 16 \text{ in } 16(c) \qquad (xxx) \qquad \text{with } x = 0.1003$$

There are two kinds of nonequivalent link, distinguished in Fig. 9.10 as A (along triad axis) and B; these are equal in length if $x = (\sqrt{2} - 1)/4 = 0.1036$. This is a uniform 6^6 net that is topologically and geometrically different from Nets 6^6-(a) and (b):

	x	y	
Net 6^6-(a) and b	12	6	
Net 6^6-(c)	9	6 (one-quarter)	$y_{\text{mean}} = 4\frac{1}{2}$
		4 (three-quarters)	

The two sets of nonequivalent links A and B have y values of 6 and 4, respectively. This net can be constructed with bond angles approximately equal to 109°, and its geometry is related in a simple way to that of the cubic and hexagonal diamond nets.

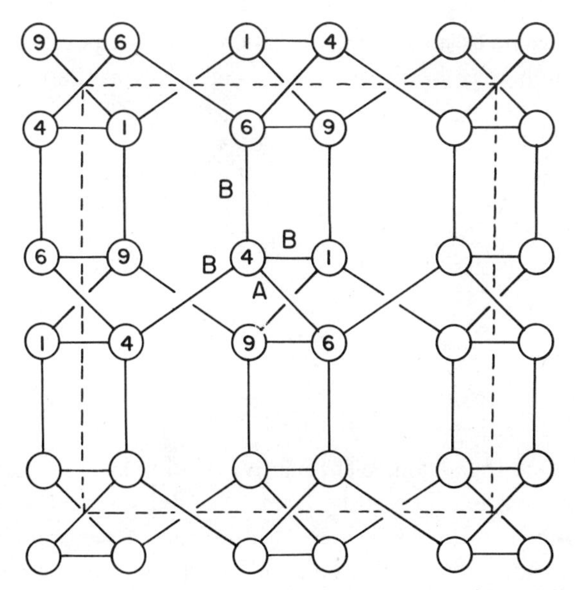

Fig. 9.10. Projection of the structure of a high-pressure form of silicon.

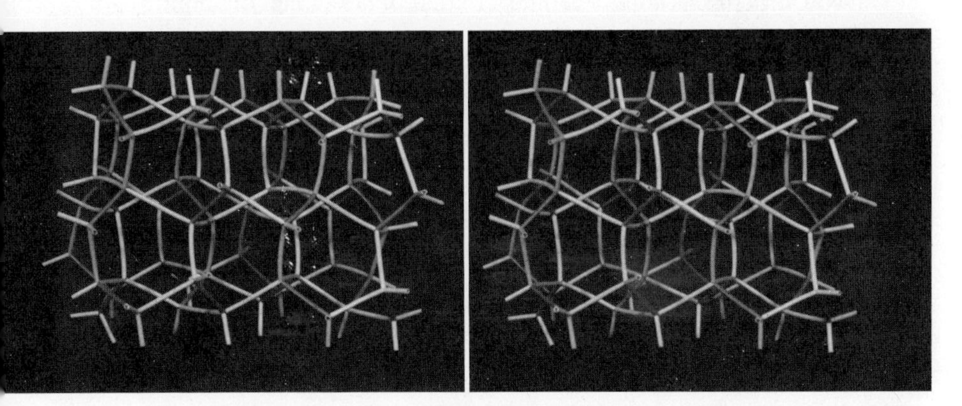
Fig. 9.11. The net representing the structure of high-pressure silicon.

The two extreme arrangements of links around the points at the ends of a particular link may be described as staggered (*trans*) or eclipsed (*cis*).

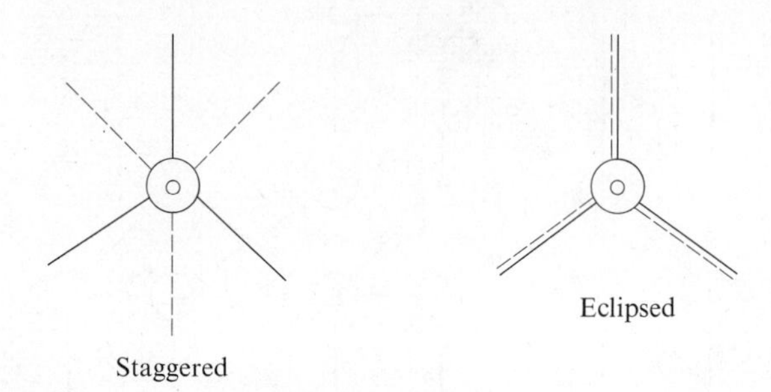

In cubic diamond the arrangement relative to all links is staggered, and the net can be constructed by starting from two points with their six terminal links so arranged, adding points in all directions, requiring only that the staggered configuration of links be maintained about every link added. In hexagonal diamond the arrangement is staggered relative to three-quarters of the links but eclipsed for the remaining one-quarter of the links. The polytypes of SiC and ZnS represent intermediate cases. In this respect the h.p. silicon structure is complementary to that of hexagonal diamond:

Related nets	Fraction of links about whose ends arrangement is "staggered"					
	0	0	$\frac{1}{4}$	$\frac{1}{2}$	$\frac{3}{4}$	All
$(6^3)(6^6)$	Gas hydrates	?	h.p. Si	?	Hexagonal diamond (Net 6)	Cubic diamond (Net 1)
					⟵	⟶
$(5^6)(5^{12})$	Radiating 5^6	Periodic 6^6 nets			SiC polytypes radiating 6^6	

No net has been found in which the fraction is $\frac{1}{2}$, and it also seems likely that there is no periodic 3D 6^6 net in which the bond arrangement about the ends of *all* links is eclipsed. This arrangement of links about successive links leads to the formation of convex 3-connected polyhedra, and there is no such polyhedron with all faces 6-gons. In fact the preferred polygonal circuit is obviously a pentagon, the internal angle of which (108°) is very close to the regular tetrahedral value. We find therefore a net 5^6 which radiates out from a central pentagonal dodecahedron (Fig. 2.5). Apparently the closest we can approach either a periodic 3D 4-connected 6^6 net with entirely eclipsed arrangements of bonds or a periodic 3D 4-connected 5^6 net is one such as the gas hydrate structure:[2]

Cubic Space group $Pm3n$ (No. 223)

$Z = 46$ in $6(c)$; $16(i)$; $24(k)$

This net can be constructed with equal bonds and with bond angles close to the regular tetrahedral value, but there are points of three kinds with point symbols 5^6, 5^56, and 5^46^2. This is a polyhedral space-filling (by pentagonal dodecahedra and 14-hedra, $f_5 = 12$, $f_6 = 2$), and the bond arrangement is eclipsed about *every* link—it is illustrated in Fig. 9.12.

The study of 6^6 nets revealed that there is an infinite net that radiates from a line (Fig. 9.13) and is closely related to the diamond and SiC polytype structures. The latter consist of slices of various thicknesses of the two diamond structures (or in the case of ZnS, of the sphalerite and wurtzite structures). This radiating net is built of layers of two kinds, alternate ones having a central plane hexagon, and comparison with elevations of cubic and hexagonal diamond shows that the structure can be described as a 6-fold rotation twin of cubic diamond, the structure along the twin planes corresponding to that of hexagonal diamond (Fig. 9.14). The arrangement of links is eclipsed about only those bonds emphasized in Fig. 9.14, and the fraction of such links falls as the net is extended outward from the central unique 6-fold axis. For cylindrical portions of this net enclosed within the

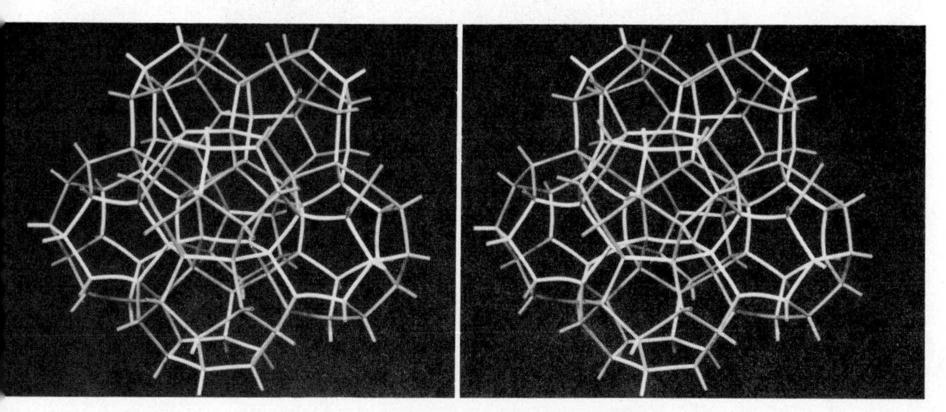

Fig. 9.12. Water framework in chlorine hydrate.

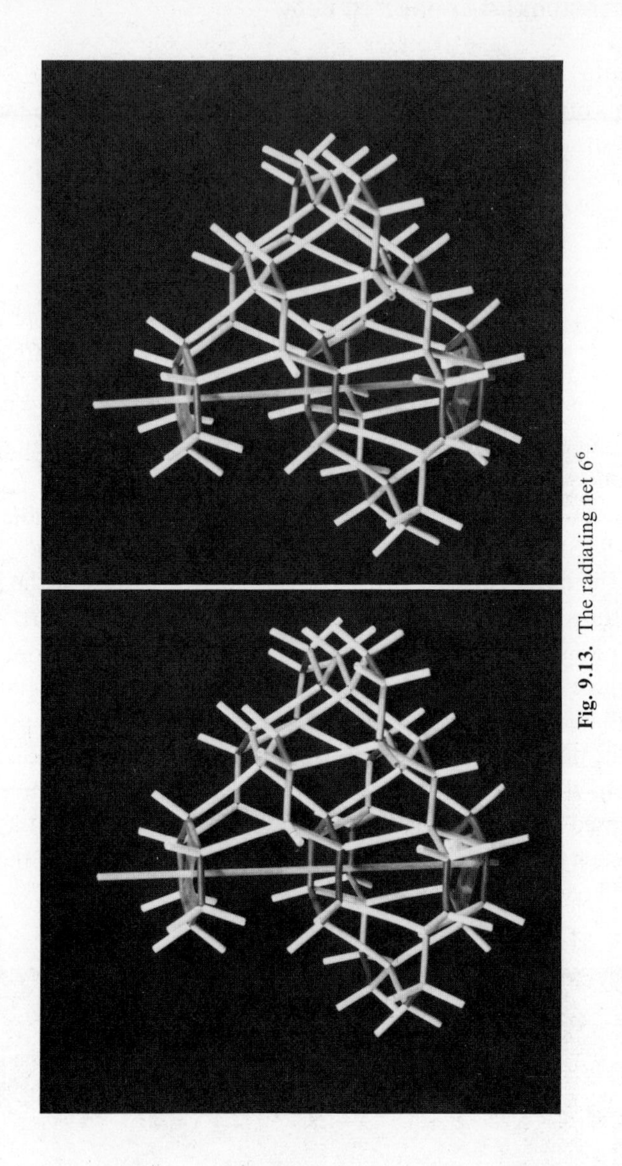

Fig. 9.13. The radiating net 6^6.

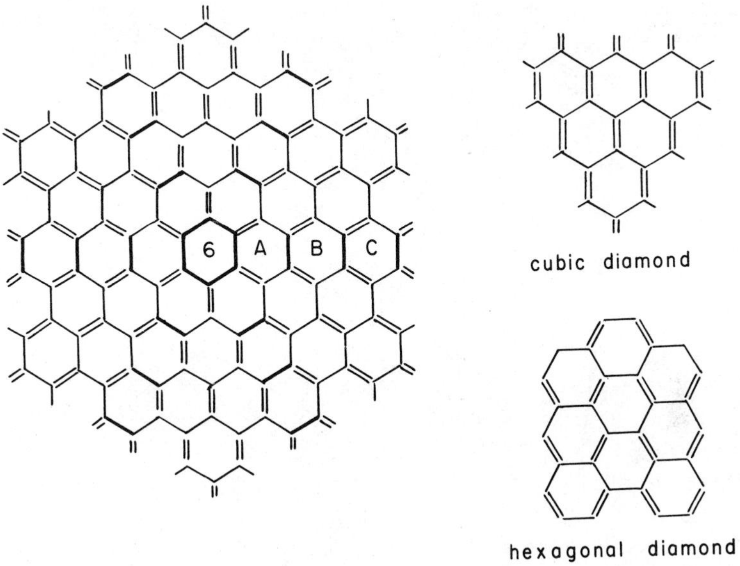

6 A B C

cubic diamond

hexagonal diamond

Fig. 9.14. The radiating 4-connected net 6^6.

limits *A*, *B*, *C*, and so on, the fractions are $\frac{1}{2}, \frac{1}{4}, \frac{1}{6}, \frac{1}{8}, \ldots$, approaching, though never reaching, the zero value characteristic of the cubic diamond net.

Nets $m.n^5$

The three simplest nets of this kind are closely related (Fig. 9.15) for they consist of sets of points generated by screw axes that are cross-connected by 3-gons or 4-gons at appropriate levels:

2_1 + 3-gons:	3.6^5	
3_1 + 3-gons:	3.7^5	(Net 2 of Table 9.2)
2_1 + 4-gons:	4.6^5-a	(Net 7 of Table 9.2)

The nets 3.6^5 and 4.6^5 are particularly closely related, since they consist of "cylindrical" tunnels on the walls of which the plane 6-gon net is inscribed, three tunnels being linked by 3-gons in 3.6^5 and four by 4-gons in 4.6^5. The latter represents the arrangement of B atoms in CrB_4, and 3.6^5 corresponds to the positions of the centers of the spheres in the open sphere packing 4_2

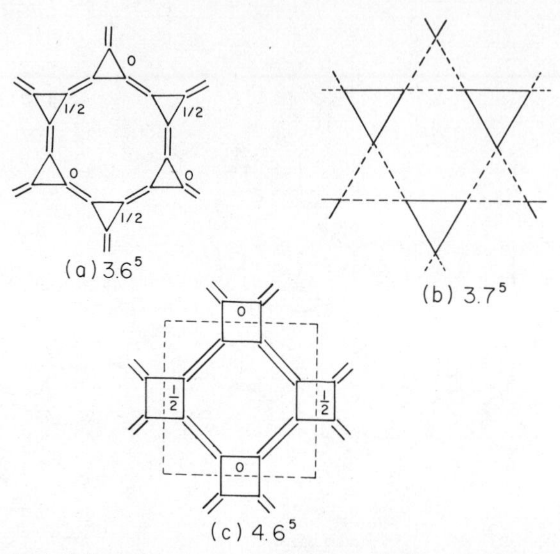

Fig. 9.15. Projections of 4-connected nets $m.n^5$. Numerals in (*b*) indicate heights of points as multiples of $c/3$.

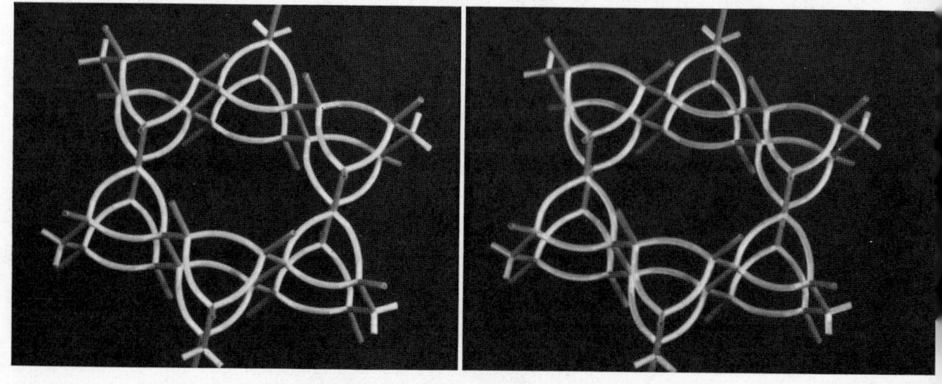

Fig. 9.16. The 4-connected net 3.6^5.

(hexagonal variant) of Heesch and Laves (see also Chapter 12). The net 3.7^5 is enantiomorphic; it is necessary that all the 3-fold axes have the same chirality.

All three nets can be constructed with equal bonds; the nets are illustrated in their most symmetrical configurations in Figs. 9.16–9.18.

As noted in Table 9.3 4.6^5 nets are now known with $Z = 4$, 8, and 24.

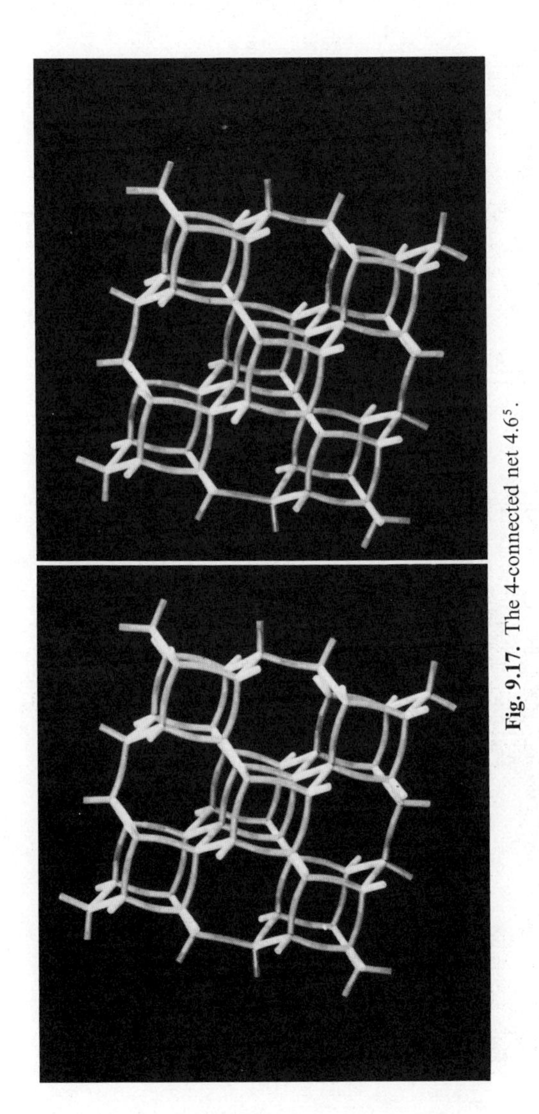

Fig. 9.17. The 4-connected net 4.6^5.

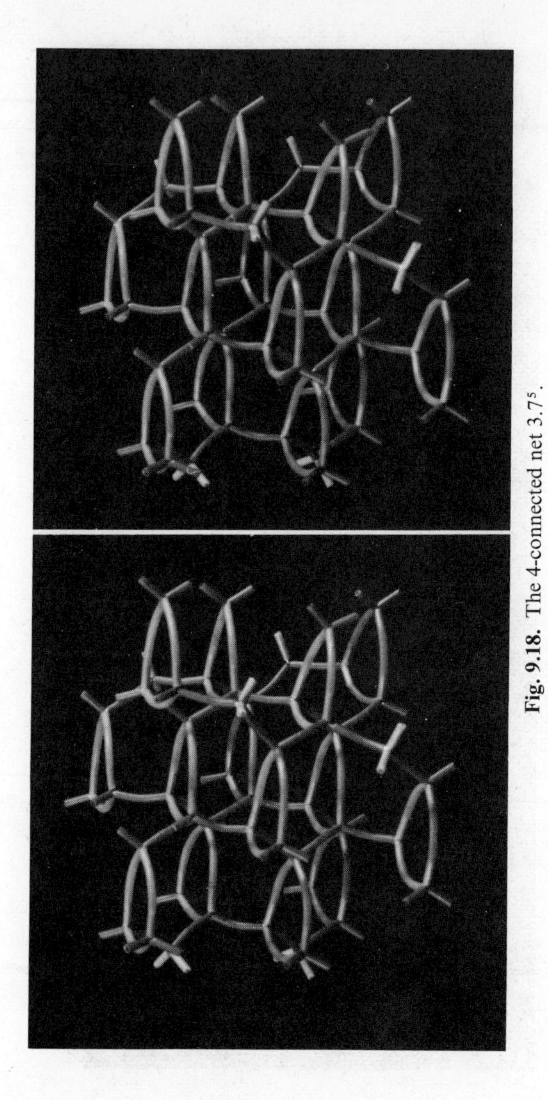

Fig. 9.18. The 4-connected net 3.7^5.

Nets m^2n^4

The six known nets of this family are the following:

$$3^210^4 \qquad 4^26^4 \text{ (two)} \qquad 4^28^4 \qquad 8^26^4 \text{ (two)}$$

Net 3^210^4

If the midpoints of the links around each point of a 3-connected polyhedron or net are joined to form triangles, the resulting 4-connected polyhedron or net contains a continuous system of vertex-sharing triangles:

tetrahedron (3, 3) $\rightarrow$ octahedron $\begin{pmatrix}3 \\ 3 \end{pmatrix}$, 4) — plane net (6, 3) $\rightarrow \left(\begin{matrix}3\\6\end{matrix}, 4\right)$

$$\text{tetrahedron } (3, 3) \rightarrow \text{octahedron } \left(\begin{matrix}3\\3\end{matrix}, 4\right) \qquad \text{plane net } (6, 3) \rightarrow \left(\begin{matrix}3\\6\end{matrix}, 4\right)$$

$$\text{cube } (4, 3) \rightarrow \text{cuboctahedron } \left(\begin{matrix}3\\4\end{matrix}, 4\right) \qquad \text{3D net } (n, 3) \rightarrow \left(\begin{matrix}3\\n\end{matrix}, 4\right)$$

$$\text{dodecahedron } (5, 3) \rightarrow \text{icosidodecahedron } \left(\begin{matrix}3\\5\end{matrix}, 4\right)$$

Since two triangles meet at each point, the most symmetrical 3D net of this type is one with the point symbol 3^2n^4 formed from a uniform 3-connected net $(n, 3)$. We may refer to the links of the original 3-connected net by the numbers labeling their midpoints, which are the points of the derived 4-connected net (Fig. 9.19). The point symbol 3^2n^4 implies that all four circuits *ac*, . . . , *ad*, . . . , *bc*, . . . , and *bd*, . . . , are n-gons. This in turn means that in the 3-connected net all four sequences of links 134, 135, 234, and 235 must be parts of n-gon circuits. This requirement, involving combinations taken three at a time of five links around two points, is much more rigorous than that for uniformity, which implies only that all combinations of three links around one point taken two at a time (namely, 12, 13, 23, and 34, 35, 45) form parts of n-gon circuits.

It appears that the only 3^2n^4 net is 3^210^4, formed from the cubic net $(10, 3)$-*a*, which is unique among 3-connected nets in not only having all points equivalent (like several uniform 3-connected nets) but also in having all links equivalent. The derived 4-connected net is described as follows:

Cubic Space group $I4_132$ (No. 214) $Z = 12$ in $12(c)$

The points of this net are the positions of the centers of the spheres in *H* and

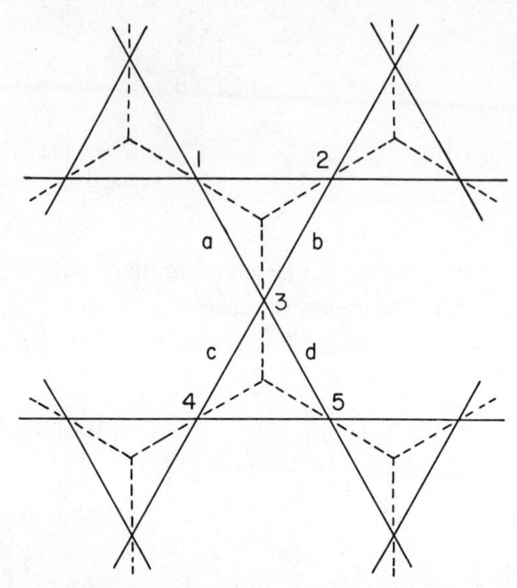

Fig. 9.19. Relation between a net n^3 and a net $3^2.n^4$.

$L\,4_3$. This net is mentioned in the Chapter on "Interpenetrating Nets" and also in Chapter 7 in connection with octahedral AX_3 structures derived from 3-connected nets.

Triangulated 4-connected nets derived from certain other uniform 3-connected nets have more complex point symbols because the points of the 4-connected net are the midpoints of *nonequivalent* links of the 3-connected net:

3-connected net	Point symbol of 4-connected net	Figure
(10, 3)-*b*	$(3^2 10^4)(3^2 10^3 11)_2$	9.24
(10, 3)-*c*	$(3^2 10^2 11^2)(3^2 10^3 11)_2$	—
(8, 3)-*b*	$(3^2 8^2 9^2)(3^2 8^3 9)_2$	9.25

Only one compound (GeS_2) is known to form a structure of this type. The positions of the Ge atoms are those of the net derived from (10, 3)-*b*, and not those of the topologically simpler net derived from (10, 3)-*a*. The unit cell of the structure of GeS_2 contains 24 Ge atoms that are in the positions 8(*a*) and 16(*b*) of the space group *Fdd*2 (No. 43). This net is illustrated as a stereo-pair in Fig. 9.24. (See note on p. 178 concerning the space group of GeS_2.)

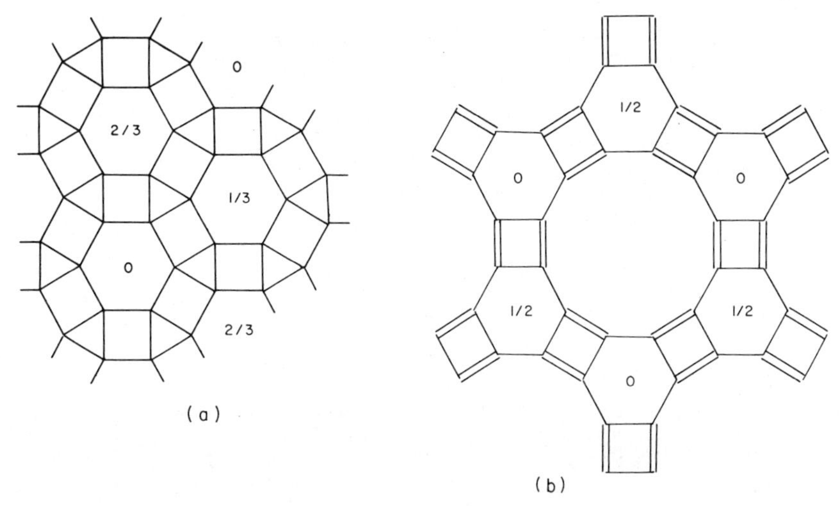

Fig. 9.20. Projections of two nets $4^2 6^4$.

Nets $4^2 6^4$

The two nets shown in projection in Fig. 9.20 are formed from points generated by 3-fold and 2-fold screw axes (perpendicular to the plane of projection) linked by plane 6-gons. In Fig. 9.20*a* axes 3_1 and 3_2 alternate; in Fig. 9.20*b* the 2_1 axes are plane zigzags. The values of Z_t are, respectively, 6 and 12. The net of Fig. 9.20*a* is better known in the form given in Fig. 9.21*a*. The links outline truncated octahedra (Fedorov's fifth space-filling parallelepiped), and the most symmetrical configuration has cubic symmetry (Table 10.1), equal links, and bond angles at each point 90° (two) and 120° (four). The points of this cubic framework represent the positions of the Si(Al) atoms in the aluminosilicate ultramarine and of the water molecules in $HPF_6.6H_2O$. A less symmetrical variant of this net [e.g., the position 12(*h*) in *Pm*3*m*, $x \neq \frac{1}{4}$] appears in Fig. 9.21*b*; if $x = \frac{1}{4}$ it becomes the net of Fig. 9.21*a*.

The net shown in projection in Fig. 9.20*b* and as a stereo-pair in Fig. 9.26 is not a polyhedral space-filling but it may be described as consisting of columns of 3-connected polyhedra ($f_4 = 6, f_6 = 5$)—compare $f_4 = 6, f_6 = 8$ for the Fedorov net—which share opposite 6-gon faces; these columns are arranged around tunnels on which the simple 6-gon net is inscribed. This description of this net can be compared with that of the $4^2 6^3 8$ net 8′, which could be described as its tetragonal analogue.

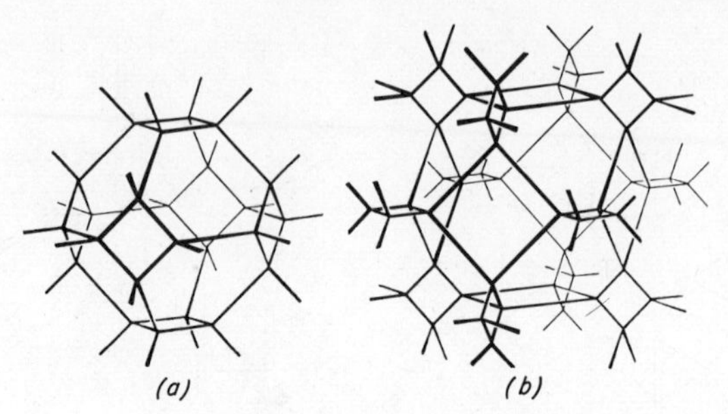

Fig. 9.21. Configurations of the cubic net $4^2 6^4$.

We refer later to two polyhedral space-fillings that are intermediate between the Fedorov net $4^2 6^4$ and the net 5^6 (known only as a radiating net); both, like the gas hydrate structure already described, have two or more kinds of nonequivalent point.

Net $4^2 8^4$

Although all points in this net (Net 15 of Table 9.2) have the same point symbol, they are of two kinds with regard to bond arrangement. In the most symmetrical configuration of this net [Fig. 9.5 (15)] the bonds around half the points are coplanar and those around the remainder are tetrahedral. These arrangements cannot, however, be simultaneously the most symmetrical possible (i.e., square coplanar and regular tetrahedral) because one of the tetrahedral angles is also the supplement of one of the "coplanar" bond angles (Fig. 9.22). A highly symmetrical form of this net represents the structure of PtS (cooperite), in which the metal atoms and the sulfur atoms form coplanar and tetrahedral bonds, respectively. The net can be referred to a

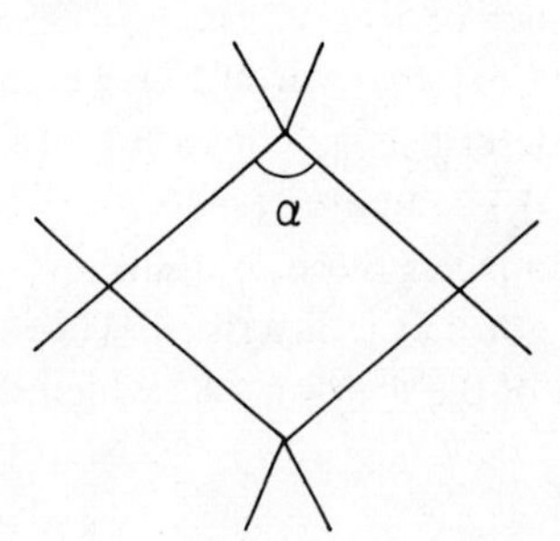

Fig. 9.22. Interdependence of bond angles (see text).

unit cell containing four points:

Tetragonal Space group $P4_2/mmc$ (D_{4h}^9) (No. 131)

$Z = 4$ in $2(c)$, $(0\frac{1}{2}0)$, $(\frac{1}{2}0\frac{1}{2})$; in $2(e)$, $(00\frac{1}{4})$, $(00\frac{3}{4})$

which has *a* axes at 45° to those of Fig. 9.5 (15).

Nets $8^2 6^4$

The two nets (Nets 4 and 5 of Table 9.2) with this point symbol are derived from the closely related Nets (8, 3)-*b* and *a*; respectively, by coalescing pairs of points—compare the projections of Fig. 9.3*e*, *f*. Net 4 represents the structure of NbO if Nb and O atoms are placed at alternate points, whereas Net 5 represents the arrangement of Si atoms in quartz. The most symmetrical configurations of these nets are the following:

Net 4: Cubic Space group $Im3m$ (O_h^9) (No. 229)

$Z = 6$ in $6(b)$, $(0\frac{1}{2}\frac{1}{2})$, and so on*

Net 5: Hexagonal Space group $P6_222(D_6^4)$ (No. 180)

$Z = 3$ in $3(c)$, $(\frac{1}{2}\frac{1}{2}\frac{1}{3})$, $(\frac{1}{2}00)$, $(0\frac{1}{2}\frac{2}{3})$

Further 4-connected nets are generated by joining the midpoints of bonds around the points of the 3-connected nets (8, 3)-*b* and *a* (Fig. 9.23); one of these has been described in the section on the net $3^2 10^4$.

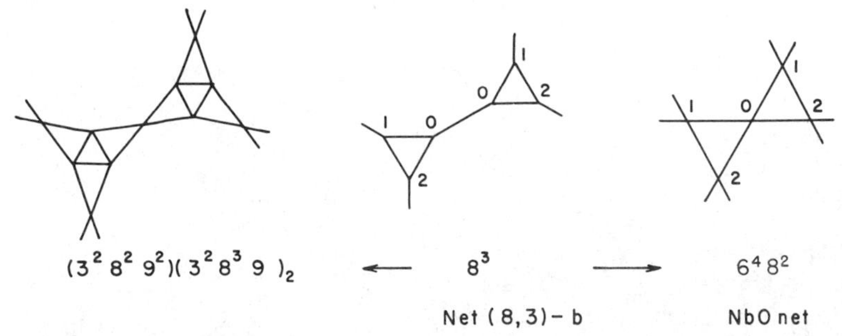

Fig. 9.23. Relations between 3- and 4-connected nets. Two 4-connected nets formed from (8, 3)-b by joining mid-points of bonds around the points of the 3-connected net (left) or by coalescing pairs of points (right).

* See note p. 110.

Nets m^3n^3

The first two nets of this type arise by placing the 3-connected regular solids with four and eight vertices at the points of uniform 4- and 8-connected 3D nets. (The net derived from the pentagonal dodecahedron would require a 20-connected net.) The first net, in which tetrahedra replace the points of the diamond net, has the symbol $3^3 12^3$ and corresponds to the positions of the centers of the spheres in *H* and *L* 4_4. The net $4^3 8^3$ arises by replacing the points of the *I* lattice by cubes. One net $4^3 6^3$ (Table 10.1) is shown in Fig. 18.5; another is the basis of the structure of the zeolite faujasite.

Reference has already been made to Figs. 9.24, 9.25, and 9.26 which are stereo-pairs, of respectively the triangulated nets derived from (10, 3)-*b* and (8, 3)-*b* and the $4^2 6^4$ net shown in projection in Fig. 9.20*b*.

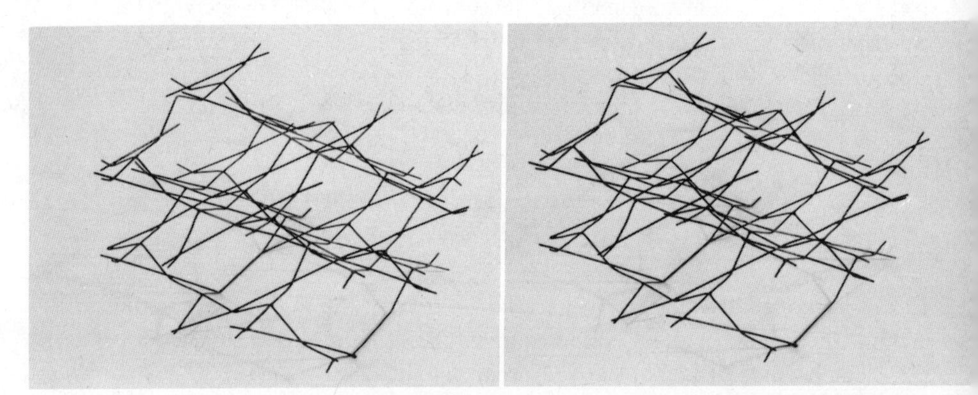

Fig. 9.24. Net representing the positions of Ge atoms in GeS_2.

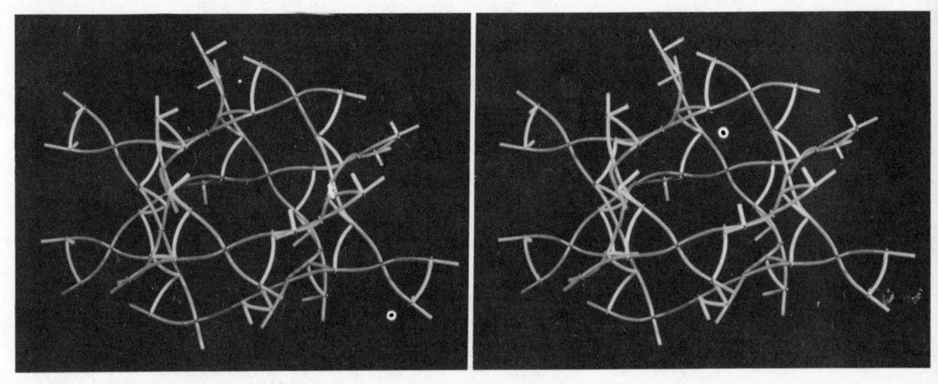

Fig. 9.25. A 4-connected net derived from the 3-connected net (8, 3)-*b*.

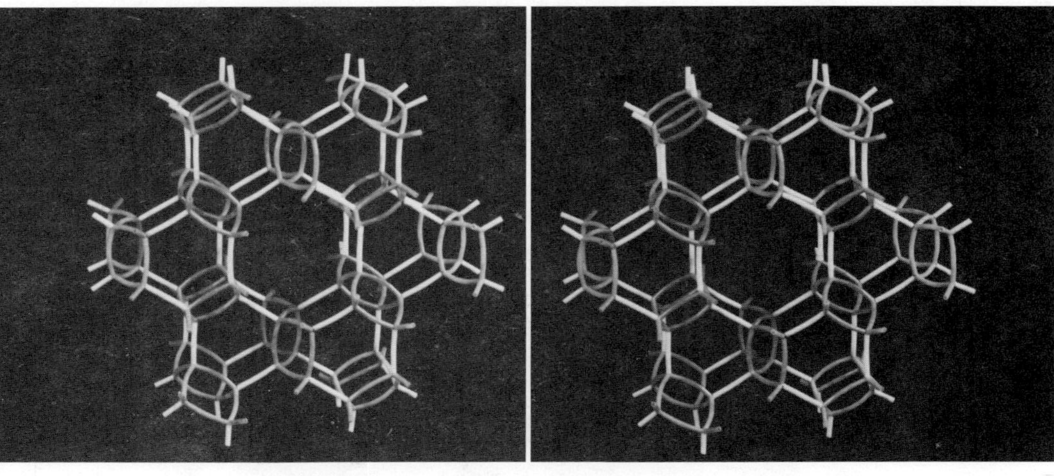

Fig. 9.26. The net $4^2 6^4$ shown in projection in Fig. 9.20*b*.

Nets $1^a m^b n^{6-a-b}$

In nets of this family all points are topologically equivalent and polygons of three kinds meet at each point. They include Net 8 of Table 9.2, a closely related tetragonal net that we call Net 8′, and one of Andreini's space-fillings (see Table 10.1):

$$\left.\begin{array}{r}\text{Net } 8\\ 8'\end{array}\right\} \quad 4^2 6^3 8$$

$$\text{Andreini, Fig. 21} \qquad 4^3 6^2 8$$

Net 8 was derived (Fig. 9.1*g*) from the plane 6-gon net, but its relation to Net 8′ is more obvious from its elevation (Fig. 9.27*b*). Both these nets are formed by 2_1 axes (plane zigzags) perpendicular to the plane of the paper and cross-connected by 8-gons. In the most symmetrical configurations of these nets (Fig. 9.28 and 9.29) the conformations of the 8-gons are chair (Net 8) and plane and boat (Net 8′). In an earlier paragraph we described a new net (Fig. 9.26) that has the same point symbol ($4^2 6^4$) as the Fedorov net and can be

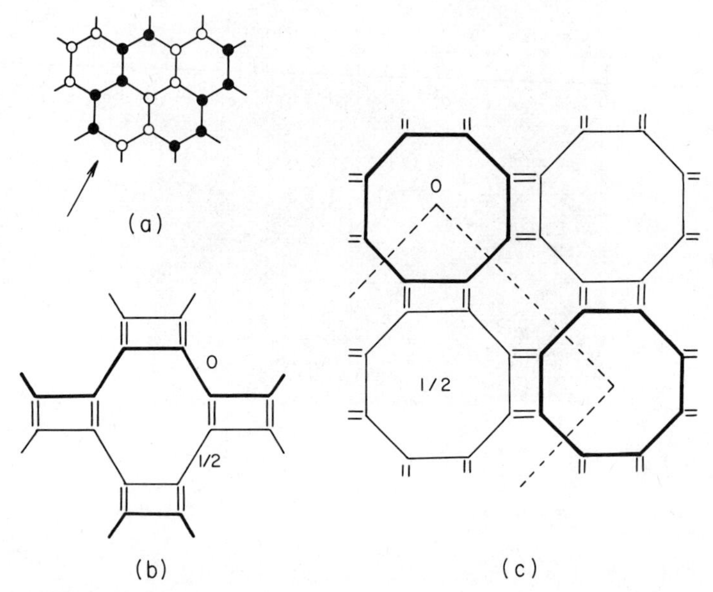

Fig. 9.27. (*a*) Plan of Net 8. (*b*) Elevation in direction of arrow in (*a*). (*c*) Net 8′. Links at heights 0 and $\frac{1}{2}$ as heavy and light lines in (*b*) and (*c*), respectively.

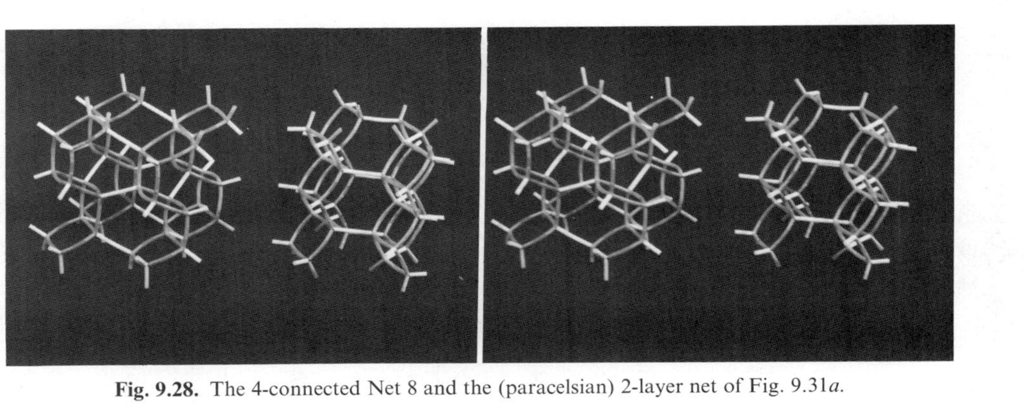

Fig. 9.28. The 4-connected Net 8 and the (paracelsian) 2-layer net of Fig. 9.31*a*.

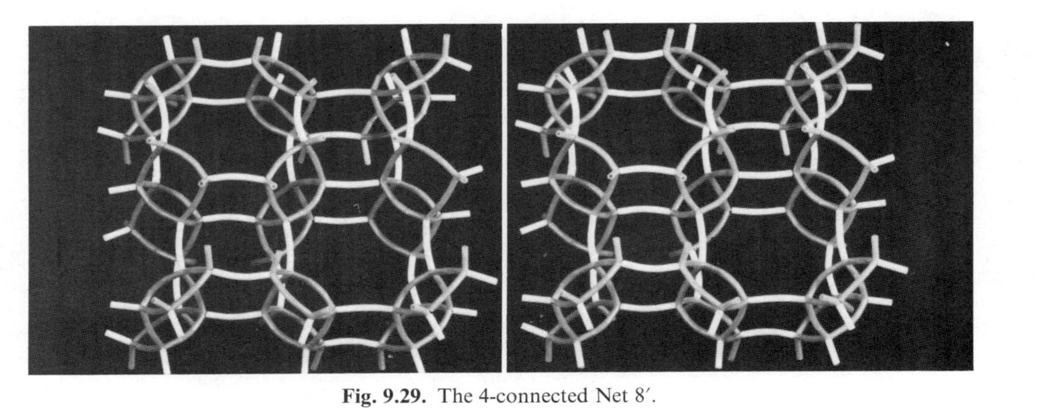

Fig. 9.29. The 4-connected Net 8′.

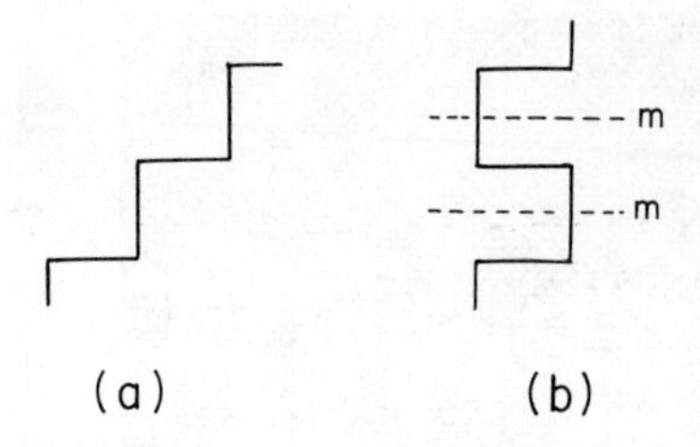

Fig. 9.30. Two types of "ladder" formed by 4-gons. Each line is an edge of a 4-gon that extends perpendicular to the plane of the paper and toward the reader.

described as containing columns of polyhedra $f_4 = 6$, $f_6 = 5$ arranged around tunnels on which the simple 6-gon net is inscribed. The Net 8′ may be described in a similar way as columns of 3-connected polyhedra $f_4 = 8$, $f_6 = 4$, $f_8 = 2$, arranged around 8-sided tunnels on which the same plane net is inscribed.

In Nets 8 and 8′ and also in the net $4^2 6^4$ of Fig. 9.26 the 4-gons form "ladders" (Fig. 9.30*a*) with their axes parallel. More buckled sequences of of 4-gons ("double crankshafts", Fig. 9.30*b*), are a feature of the structures of felspars. Frameworks containing both these types of subunit can be derived from plane nets containing 4-gons, and the simplest arise from the plane 4.8^2 net. In Fig. 9.31 the solid and open circles represent points to be joined downward or upward, respectively to points of adjacent layers. If the pattern of solid circles in a layer is the same as that of the open circles (Fig. 9.31*a*), then for the layers to be joined together, it is only necessary to translate each one relative to its neighbors. The number of points in the repeat unit of the 3D net is the same as in a layer, and the layer of Fig. 9.31*a* leads to "ladders" of the type of Fig. 9.30*a*. Alternatively adjacent layers may be related by a mirror plane, when the chains have the more buckled configuration (Fig. 9.30*b*), and Z has twice the value corresponding to the plane net. This latter (2-layer repeat) is the only possibility if the pattern of solid circles is not the same as that of the open circles, as is the case in Fig. 9.31*b*.

The solid and open circles of Fig. 9.31 could also represent centers of AX_4 tetrahedra having the fourth vertex pointing downward or upward, this vertex to be shared with appropriate vertices of adjacent layers. An X atom along each link is shared between two tetrahedra, resulting in the formation of a 3D structure of composition AX_2. All the structures of Fig. 9.31 can be built from tetrahedra. The 1-layer and 2-layer structures (Fig. 9.31*a*) are presented in stereoscopic form in Figs. 9.32 and 9.33. The latter represents the structure of paracelsian,[3] $BaAl_2Si_2O_8$ and of the isostructural danburite, $CaB_2Si_2O_8$.[4] The structures of felspars are based on the more complex net

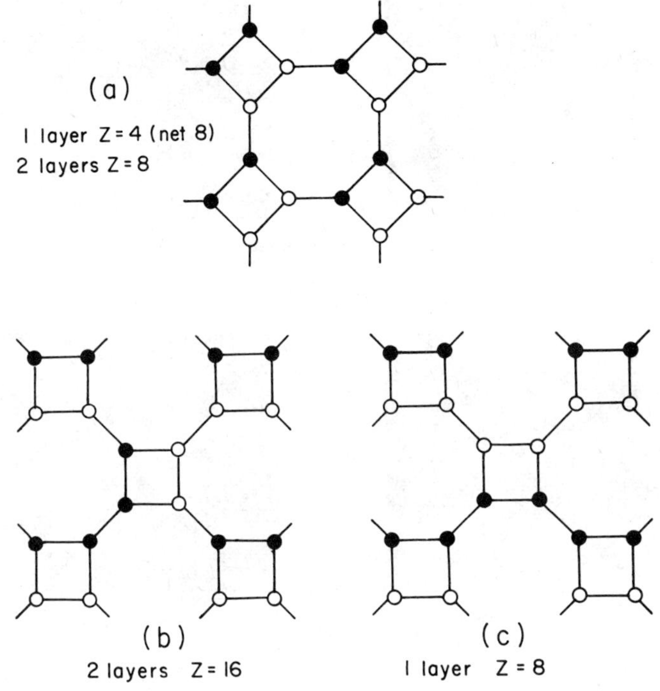

Fig. 9.31. Derivation of the felspar and related nets from the plane 4.8^2 net.

corresponding to Fig. 9.31*b*, and those of certain zeolites are based on nets derived from layers of the same general type as, for example, gismondine, $CaAl_2Si_2O_8.4H_2O$[5] on the layer (Fig. 9.31*c*). The characteristic felspar chains can also be distinguished in the structure of coesite,[6] a high-pressure form of silica. An AX_2 structure can be built from tetrahedral AX_4 groups, but this does not necessarily mean that the basic net of A atoms can be constructed with approximately linear links and angles close to $109\frac{1}{2}°$ (as in the models of Figs. 9.28 and 9.29). The possibility of noncollinear A–X–A bonds (equivalent to bent links in the basic net of A atoms) introduces a degree of freedom that permits the modification of a particular framework to accommodate ions of various sizes, as in the felspars.

The 4-connected net corresponding to the 2-layer structure of Fig. 9.31*a* has been included with the 1-layer structure (Net 8) in Fig. 9.28. The two nets in this Figure therefore correspond to the tetrahedral structures of Figs. 9.32 and 9.33.

Fig. 9.32. Tetrahedral AX_2 structure based on Net 8.

Fig. 9.33. Tetrahedral AX_2 structure based on the (paracelsian) 2-layer net of Fig. 9.31*a* (cf. Fig. 9.28).

The space-filling $4^3 6^2 8$ of Andreini's Fig. 21 is illustrated in Fig. 18.4 in the Chapter on complementary 3D polyhedra. Although not a net of the type we are considering here, the space-filling Fig. 23 of Andreini may be mentioned here. All points have the same symbol, but it is of a more complex type $(3.4.6^2 8^2)$; it is included in Table 10.1. The points of this net are the positions of the sphere centers in *H* and *L* 4_2 (cubic variant).

Nets with nonequivalent points

Under this heading we wish to mention three small groups of nets; the first includes Nets 3, 13, and 18 of Table 9.2. The similarity between the first two is indicated by their point symbols:

$$\text{Net 3:} \quad (4^2 6^2 8^2)(46^4 8)_2$$

$$\text{Net 13:} \quad (4^2 6^2 8^2)(4^2 68^3)$$

which show that one-third of the points in Net 3 have the same point symbol

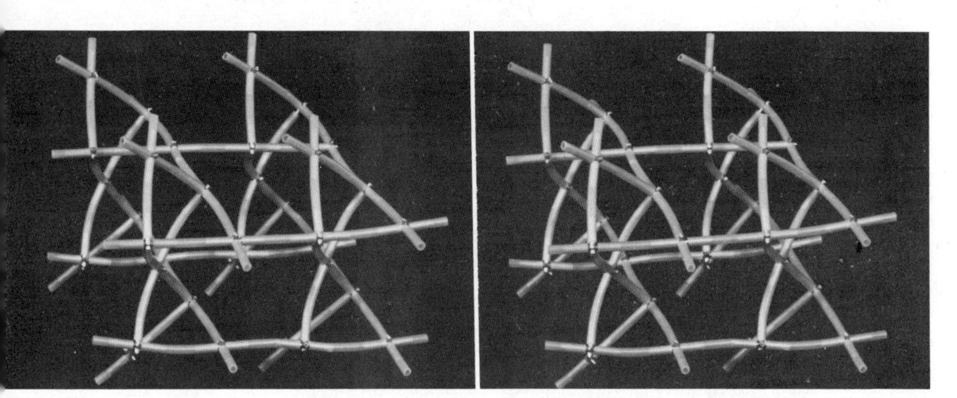

Fig. 9.34. The 4-connected Net 18.

as one-half of the points in Net 13. In Net 18 there are three kinds of non-equivalent point:

$$\text{Net 18:} \qquad (4.8^5)(4^2 8^4)(4^3 8^3)_2$$

of which one-quarter have the same symbol as all the points of Net 15. Net 18 is illustrated in Fig. 9.34.

The two nets of our second group are the edges of polyhedral space-fillings that are related both to the Fedorov space-filling by truncated octahedra and also to the gas hydrate and radiating dodecahedral nets described in the section on 6^6 nets; the gas hydrate structure belongs properly in the present section. The 14-hedron of Lord Kelvin is the truncated octahedron, modified by curvature of edges and faces to give interbond angles of $109\frac{1}{2}^\circ$ instead of 90° (two) and 120° (four) in the resulting 3D 4-connected net. It has been shown recently[7] that two other 14-hedra can be derived from the Kelvin polyhedron and that one of them approximates much more closely than the Kelvin 14-hedron to the distribution of 4-, 5-, and 6-gons found in aggregates of soap bubbles, biological cells, and metal crystal grains. The first of these (Fig. 9.35*b*) is derived from the Kelvin 14-hedron (Fig. 9.35*a*) by taking any edge common to two hexagons together with the edges that meet at each end of this edge, rotating the whole group through 90° and reconnecting the edges. The same operation performed with the similar group of edges on the opposite side of the polyhedron converts Fig. 9.35*b* to *c*. All three polyhedra of Fig. 9.35 have the same numbers of faces (14), vertices (24), and edges (36), and the angles between the edges remain $109\frac{1}{2}^\circ$. The faces are of the following kinds:

	f_4	f_5	f_6
Fig. 9.35*a*	6	0	8
Fig. 9.35*b*	4	4	6
Fig. 9.35*c*	2	8	4

(The next member of this family is the polyhedron $f_5 = 12$, $f_6 = 2$, which occurs together with the pentagonal dodecahedron in the gas hydrate structure.) The positions of the vertices in the 14-hedra packings are

Fig. 9.35*a* 12(*d*) in space group *Im*3*m* (No. 229)

Fig. 9.35*c* 4(*d*) and 8(*j*) in space group $P4_2/mnm$ (No. 136)

All three polyhedra pack to fill space, though only in the space-filling by Fig. 9.35*a* are all the polyhedra similarly oriented (Fedorov's restriction). In the Fedorov space-filling all points (vertices) are equivalent, but in the space-fillings by Fig. 9.35*b*, *c* there are, respectively, three and two kinds of non-equivalent points. The point symbols of the resulting 3D 4-connected nets show their relationship to other 4-connected nets:

Figure	Net	Types of point in 3D 4-connected net				
9.35*a*	Fedorov space-filling	4^26^4				
9.35*b*	Williams space-filling	4^26^4	4.5^36^2	5^46^2		
9.35*c*	Williams space-filling		4.5^46	5^46^2		
9.12	Gas hydrate net			5^46^2	5^56	5^6
2.5	Radiating dodecahedral net					5^6

The space-fillings of Fig. 9.35*a* and *c* are illustrated in Figs. 9.36 and 9.37.

Last we include here two nets in which the shortest circuits are 5-, 6-, 7-, and 8-gons. The projection of Fig. 9.38 is that of a high-pressure form of Ge,[1] and it also represents the positions of Si atoms in a high-pressure polymorph of SiO_2, keatite,[8] and of the O atoms in ice-III.[9] The tetragonal form of this (enantiomorphic) net is described as follows:

Tetragonal Space group $P4_12_12$ (D_4^4) (No. 92) or

$P4_32_12$ (D_4^8) (No. 96)

$Z = 12$ in 4(*a*), $xx0$, etc.; $x = 0.075$

8(*b*), $x'yz$, etc.; $x' = \frac{1}{6}$, $y = \frac{3}{8}$, $z = \frac{1}{4}$

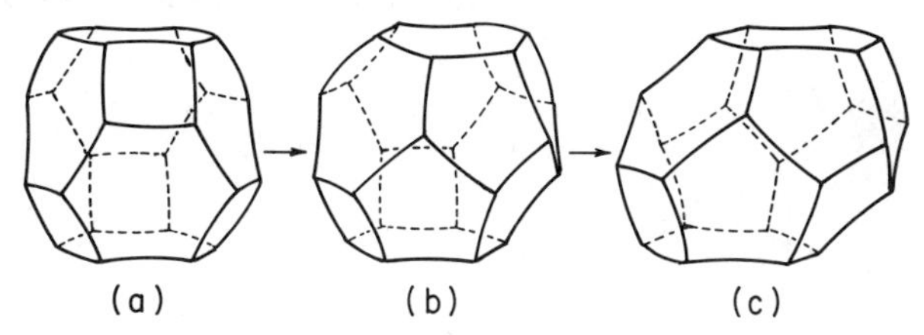

Fig. 9.35. The transformation from the α-tetrakaidecahedron (*a*) through a polyhedron with four 4-gon, four 5-gon, and six 6-gon faces (*b*) to the β-tetrakaidecahedron (*c*). (After R. Williams.)

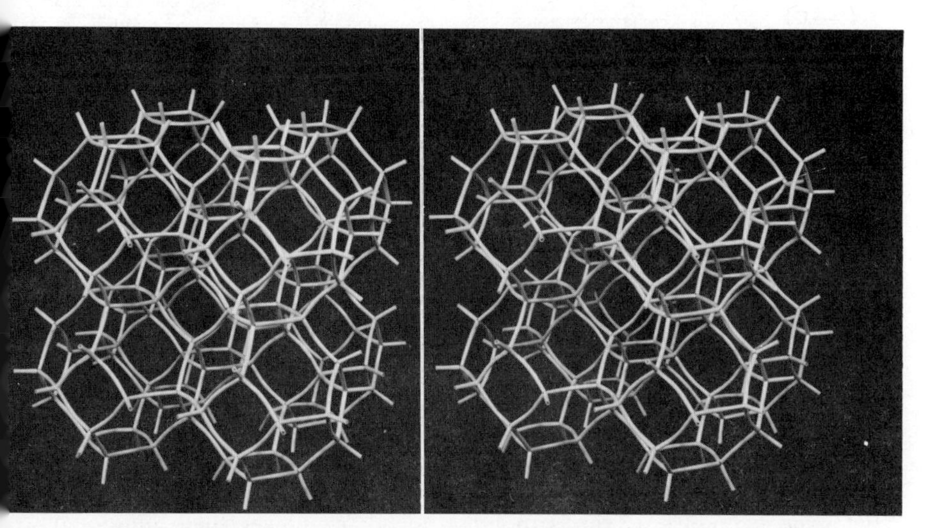

Fig. 9.36. The Kelvin form of Fedorov's space-filling arrangement of truncated octahedra.

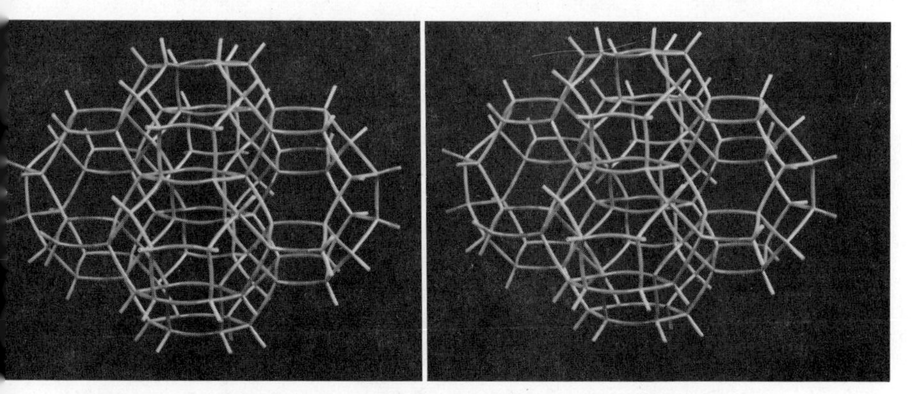

Fig. 9.37. The space-filling by the 14-hedron of Fig. 9.35*c*. (R. Williams.)

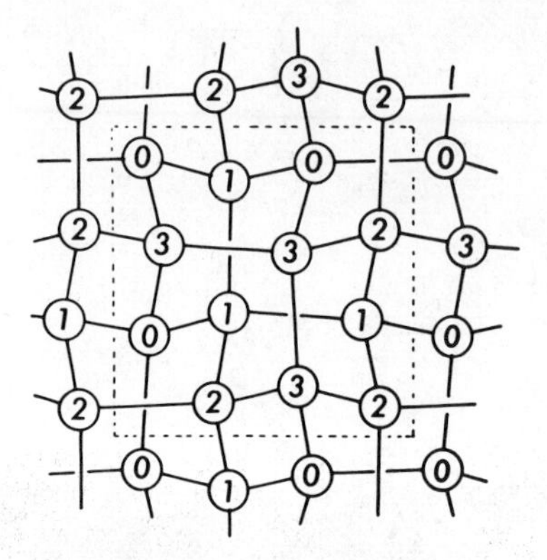

Fig. 9.38. The net that forms the basis of the structures of a high-pressure form of germanium, keatite, and ice-III. The net is projected on (001) and numbers indicate heights of atoms as multiples of $c/4$.

(The values of the parameters are those for Ge, for which $c/a = 1.177$.) The topological symbol for the net is $(5^3 7.8^2)(5^3 6.7^2)_2$.

It might be expected that the plane (3, 4)-connected 5-gon net of Fig. 14.6*b* would give rise to 3D nets in which 5-gon circuits are prominent. This plane net can indeed be distinguished in a number of crystal structures,[10] including those of PdP_2, PdPS, PdS_2 and FeS_2. In PdP_2 each 3-connected point (P atom) of the plane 5-gon net forms a fourth bond to a similar point of a layer above *or* below, so that a 3D 4-connected net is formed (Fig. 9.39). As in the PtS structure (Net 15), the bonds from the metal atom are square coplanar and those from the nonmetal atom are tetrahedral. The complete symbol of the net is $(5^4 8^2)(5^3 6^2 7)_2$. Linking of one-half of the 3-connected points in the plane 5-gon layer to similar points in an adjacent layer, these additional links being all to the same side of the layer, gives the double-layer structure of PdPS.[10]

The plane 5-gon net represents the structure of PdS_2, the 3-connected points being S atoms and the 4-connected points Pd atoms. If each 4-connected point is connected to a 3-connected point of each adjacent layer, it becomes 6-connected (octahedral) and the remaining points become 4-connected (tetrahedral). This (4, 6)-connected net represents the pyrite (FeS_2) structure. In this structure (Fig. 9.40) all the shortest circuits are 5-gons, and the symbols of the points are 5^6 (4-connected) and 5^{12} (6-connected). This structure is of special interest not only as the simplest periodic 3D net built of pentagons but also because it is an AX_2 structure built from octahedral AX_6 groups each vertex of which is common to *three* octahedra. From the geometry of the octahedron it is readily shown[11] that such a vertex-sharing structure (in which *only* vertices are shared) is possible for *regular* octahedra

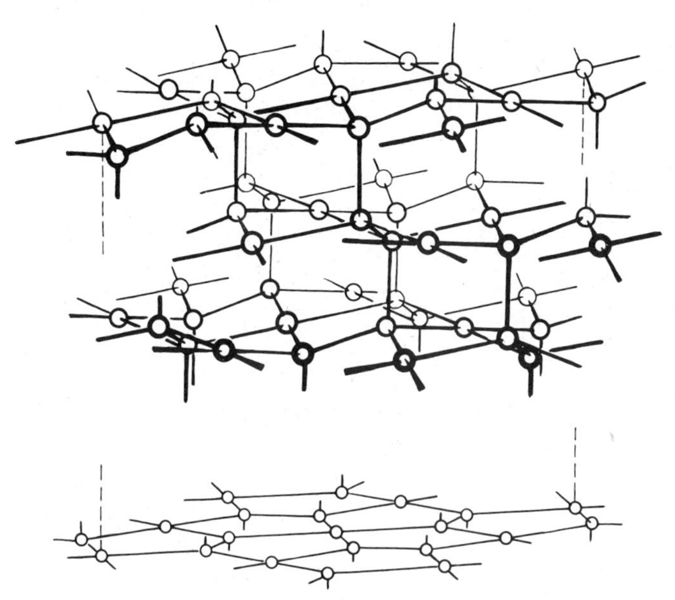

Fig. 9.39. The crystal structure of PdP_2.

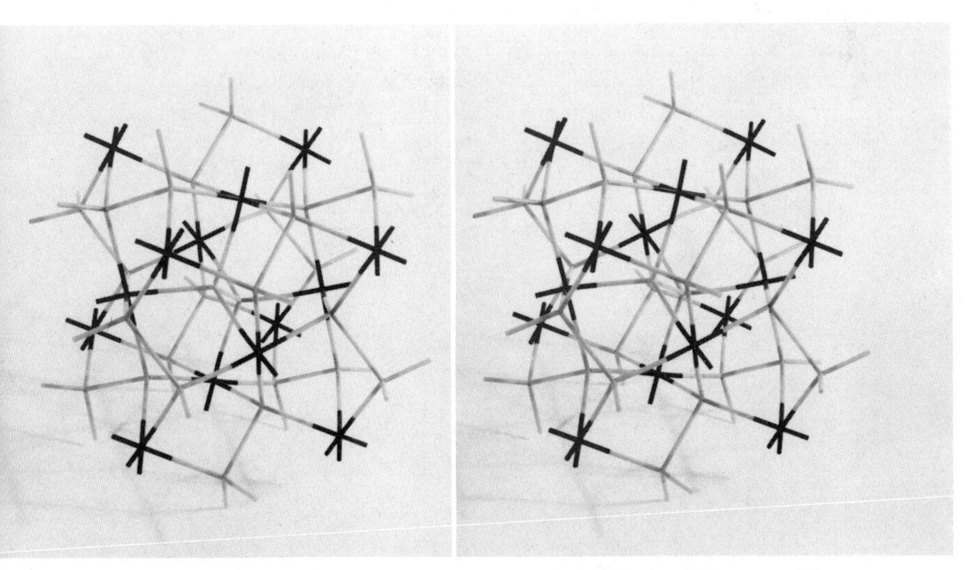

Fig. 9.40. The pyrite (FeS_2) structure, a (4, 6)-connected net that is a 3D system of 5-gons.

only if (nonbonding) distances between X atoms of different octahedra are shorter than the octahedron edge. In the pyrite structure such distances are shorter because they are the lengths of S–S bonds of S_2 groups. In its most symmetrical configuration (cubic, space group *Pa*3) the pyrite net can be constructed with equal bonds (with the parameter $u = \frac{1}{2} - (\sqrt{10} - 1)/18 = 0.3799$), but there cannot at the same time be the most symmetrical arrangement of links at each point, that is, bond angles of 90° and $109\frac{1}{2}°$.

There is an interesting topological relation between this structure and that of the high-temperature (cubic) form of SiP_2O_7.[12] The structure of the latter arises by placing an O atom on *each* link of the pyrite structure (which is that of SiP_2), giving 6-coordinated Si and pyrophosphate groups P_2O_7 consisting of two PO_4 groups with a common vertex.

References

1. J. S. Kasper and S. M. Richards, *Acta Crystallogr.*, 1964, **17**, 752.
2. K. W. Allen, *J. Chem. Soc.*, 1959, 4131; R. K. McMullan and G. A. Jeffrey, *J. Chem. Phys.*, 1965, **42**, 2725.
3. J. V. Smith, *Acta Crystallogr.*, 1953, **6**, 613.
4. V. V. Bakakin, V. B. Kravchenko, and N. V. Belov, *Dokl. Akad. Nauk SSSR*, 1959, **129**, 420.
5. K. Fischer, *Amer. Mineral.*, 1963, **48**, 664.
6. H. D. Megaw, *Acta Crystallogr.*, 1970, **B26**, 261.
7. R. E. Williams, *Natural Structure*, Eudaemon Press, 1972, p. 158.
8. J. Shropshire, P. P. Keat, and P. A. Vaughan, *Z. Kristallogr.*, 1959, **112**, 409.
9. W. B. Kamb and A. Prakash, *Acta Crystallogr.*, 1968, **B24**, 1317.
10. W. Jeitschko, *Acta Crystallogr.*, 1974, **B30**, 2565.
11. A. F. Wells, *J. Solid State Chem.*, 1973, **6**, 469.
12. E. Tillmanns, W. Gebert, and W. H. Baur, *J. Solid State Chem.*, 1973, **7**, 69.

10

Some nets with cubic symmetry

It is not our intention to deal systematically with more highly connected nets, but in the course of this work a list has been made of nets with cubic symmetry, including a number with $p > 4$ (Table 10.1).

In earlier sections we listed five 3-connected nets with cubic symmetry, namely, (10, 3)-*a* and *g* (Table 5.1) and the nets 3.20^2, 4.12^2, and 6.8^2 (Chapter 6). The (4, 6)-connected pyrite net, $(5^6)_2 5^{12}$, and the 5-connected net 5^9 of Fig. 16.27 are examples of 3D systems of pentagons that in their most symmetrical configurations also have cubic symmetry. A very simple hexagonal 5-connected net is illustrated in Fig. 15.1*e* and also in Fig. 11.13 as one of a pair of interpenetrating nets.

In some of the nets of Table 10.1 the links are the edges of space-filling polyhedra. Andreini examined a special kind of space-filling, namely, by regular or Archimedean solids, alone or in combination. Table 10.1 includes all Andreini's space-fillings that have cubic symmetry; the numbers of the figures in Andreini's paper are given, and his symbols are retained: tetrahedron (*t*), truncated tetrahedron (*tt*), cube (*c*), truncated cube (*ct*), octahedron (*o*), truncated octahedron (*ot*), cuboctahedron (*co*), truncated cuboctahedron (*cot*), rhombicuboctahedron (*rco*), and octagonal prism (p_8). Nets not illustrated elsewhere are shown in Figs. 10.1 and 10.2, and the following notes refer to particular nets as indicated in the Table.

Table 10.1 **Some nets with cubic symmetry**

p	Space group	Equivalent position	Coordinates	Point symbol	Remarks
4	*Im3m*	6(*b*)	—	$6^4 8^2$	NbO net
	Fd3m	8(*a*)	—	6^6	Diamond (H and L 4_1)
	$I4_13$	12(*c*)	—	$3^2 10^4$	H and L 4_3
	Im3m	12(*d*)	—	$4^2 6^4$	Andreini† Fig. 14 (*ot*)
	Ia3d	24(*c*)	—	$3^2 10^4$	See note 1, text
	Pm3m	24(*k*)	$y = 0.369$ $z = 0.185$	$4^3 6^2 8$	Andreini Fig. 21 (*c*, *ot*, *cot*)
	Fd3m	32(*e*)*	$x = 0.069$	$3^3 12^3$	H and L 4_4; see note 2
	Pm3m	48(*n*)	$x = 0.256$ $y = 0.104$ $z = 0.397$	$4^3 6^3$	Andreini Fig. 24′ (p_8, *cot*)
	Fm3m	96(*k*)	$x = 0.193$ $z = 0.079$	$3.4.6^2 8^2$	H and L 4_2 (cubic) Andreini Fig. 23 (*tt*, *ct*, *cot*)
5	*Pm3m*	6(*e*)	$x = 0.293$		Andreini Fig. 17 (*o*, *ct*); see note 3
	Pm3m	24(*m*)	$x = 0.138$ $z = 0.362$		Andreini Fig. 22 (*c*, p_8, *ct*, *rco*); see note 4
	Fm3m	48(*i*)	$x = \frac{1}{6}$		Andreini Fig. 24 (*tt*, *co*, *ot*); see note 5
6	*P* lattice	—	—		Andreini Fig. 5 (*c*)
	$P4_13$	4(*a*)	—		
	Fd3m	16(*c*)	—		Andreini Fig. 15 (*t*, *tt*); see note 2
	Pm3m	12(*j*)	$x = 0.207$		Andreini Fig. 20 (*c*, *co*, *rco*)
	Fm3m	32(*f*)	$x = 0.1464$		Andreini Fig. 19 (*t*, *c*, *rco*); see note 6
8	*I* lattice	—	—		—
	Pm3m	3(*c*)	—		Andreini Fig. 18 (*o*, *co*); see note 3
	$I\bar{4}3d$	12(*a*)	—		—
12	*F* lattice	—	—		Andreini Fig. 12 (*t*, *o*)

* Origin at $(\frac{1}{8}, \frac{1}{8}, \frac{1}{8})$ from center of symmetry.
† A. Andreini, *Mem. Soc. Ital. Sci.*, 1907 (3), **14**, 75.

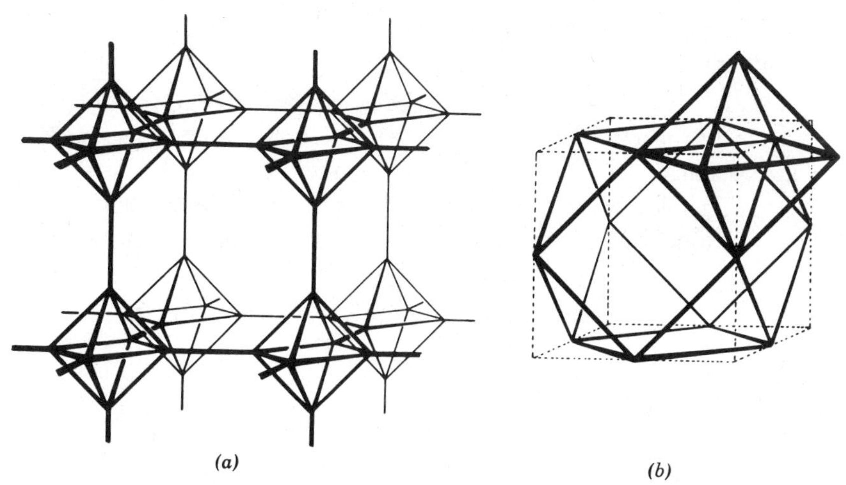

Fig. 10.1. Nets corresponding to Andreini's space-fillings Figs. 17 and 18 (see Table 10.1).

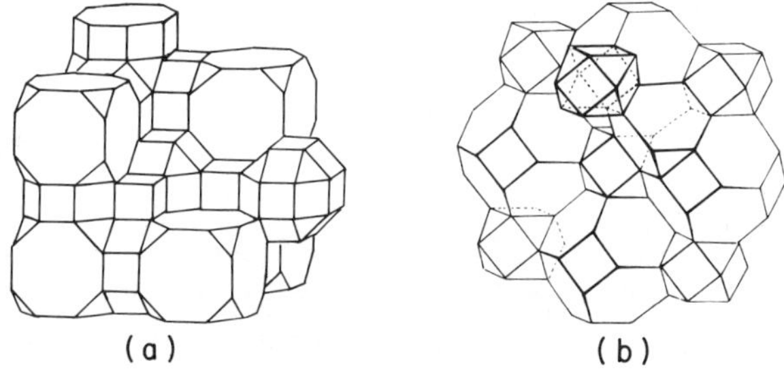

Fig. 10.2. Nets (5-connected) corresponding to Andreini's space-fillings, Figs. 22 and 24 (see Table 10.1).

1. The net $I4_13$ 12(*c*) is enantiomorphic. The net *Ia*3*d* 24(*c*) consists of two interpenetrating but independent nets of this kind of opposite chirality. It may be described as a 3D racemate. The net *Ia*3*d*, 16(*b*), bears exactly the same relation to $I4_13$, 8(*a*), which is the 3-connected net (10, 3)-*a*.

2. Tetrahedra may be placed at the nodes of the diamond net to give the 4-connected net *Fd*3*m*, 32(*e*), in which there are discrete tetrahedra, two points being placed on each link, or the 6-connected net *Fd*3*m*, 16(*c*), by joining midpoints of adjacent links. Of these two nets the former corresponds to the open packing 4_4 of Heesch and Laves, and the latter to

Andreini's packing of equal numbers of tetrahedra and truncated tetrahedra.

3. Octahedra may be placed at the nodes of a primitive cubic lattice to give the 5-connected net of Fig. 10.1*a*, which represents Andreini's packing of octahedra and truncated cubes, or the 8-connected net of Fig. 10.1*b*, which is a packing of equal numbers of octahedra and cuboctahedra. The former represents the boron framework in the CaB_6 structure.

4. This 5-connected net is illustrated in Fig. 10.2*a*.

5. This net (Fig. 10.2*b*) represents the boron framework in UB_{12}.

6. The points of this net are the positions of the O atoms of the cation framework in $(Ag_7O_8)^+NO_3^-$.

11

Interpenetrating three-dimensional nets

Any structure in which we can recognize two or more nets that cannot be separated without breaking links can be described as consisting of interpenetrating nets. The component nets may be identical or of two or more kinds. This subject is of interest for a number of reasons. First, a number of crystal structures consist of two or more interpenetrating nets; they are described later in this chapter and summarized in Table 11.3. In addition to providing an elegant way of describing certain known structures, the study of interpenetrating nets suggests types of structure that are at present unknown; indeed, the author's interest in this topic arose from the unusual X-ray diffraction effects from a crystal whose structure is still undetermined (Part 5). Second, the relation of certain nets to the closest packing of equal spheres is expected to be relevant to the adoption of particular structures by compounds A_mX_n. Consider, for example, the structure of P_2O_5 based on PO_4 tetrahedra at the nodes of the net (10, 3)-*b*, in which each tetrahedral group shares three O atoms with other similar groups, or the structure of B_2O_3 based on the net (10, 3)-*c*, in which planar BO_3 groups share all three O atoms. Presumably one factor influencing the choice of a particular 3-connected net in such a structure is the possibility of attaining a satisfactory packing of the O atoms, which occupy the greater part of the volume. If all the PO_4 tetrahedra or all the BO_3 groups are similarly oriented, not only would the P (or B) atoms be arranged at the nodes of the 3-connected net, but so also would be the sets of four (or three) O atoms. If we suppose that the O atoms would tend to be close packed, as in the structures of many crystalline metal oxides, we can approach the problem in another way, namely, by considering whether various kinds of closest packing can be broken down into a number of identical interpenetrating nets. Our third reason for studying systems of interpenetrating nets is the connection of this problem with complementary polyhedra. Although we do not deal with 3D polyhedra and their complements until later, we include the section on complementary nets here because it is so closely connected with other aspects of interpenetrating nets.

One approach is to start with a particular net and examine how two or more identical nets can interpenetrate, subject to certain obvious provisos (e.g., that points of different nets must not coincide and that links of different nets must not intersect). We could also introduce a purely geometrical restriction, that in the system of interpenetrating nets a point has no additional neighbors as near as those to which it is linked in its own net; that is, the coordination number is not increased when the nets interpenetrate. However this condition would be meaningful only if we assume all links in each net to have the same length. In the actual examples of structures consisting of interpenetrating nets, we are not concerned with nets as simple systems of linked points but as frameworks representing the basic topology of more complex structures. The points may represent, for example, the Si atoms of SiO_4 tetrahedra or Cu atoms joined through very long ligands [as in Cu(adiponitrile)$_2$$NO_3$]; thus comparison of the distances between points of different nets with distances between points within a single net is no longer relevant. An alternative approach is to start with certain arrangements of points and see how they can be broken down into sets of points, each set forming a 3D net of some kind. The following examples illustrate the two methods.

If two identical 3-connected nets (10, 3)-*a* interpenetrate as shown in Fig. 11.1*a*, related by a translation of $a/2$, each point becomes equidistant from six others. On the other hand, a D and an L form of this (enantiomorphic) net can interpenetrate without increasing the number of nearest neighbors, suggesting an interesting type of structure, at present unknown, a 3D racemate (Fig. 11.1*b*). Two interpenetrating 6.10^2 nets (Fig. 6.13) form the basis of the structure of β-quinol. The 8-connected *I* net is formed by the interpenetration of two *P* nets (6-connected). The second procedure is illustrated by the breakdown of the *I* net into two diamond nets (Fig. 11.2*a*), the *F* net into two 'extended-diamond' nets (diamond nets elongated in one direction to give two interbond angles of 90° and four of 120°; Fig. 11.2*b*), and the *P* net into

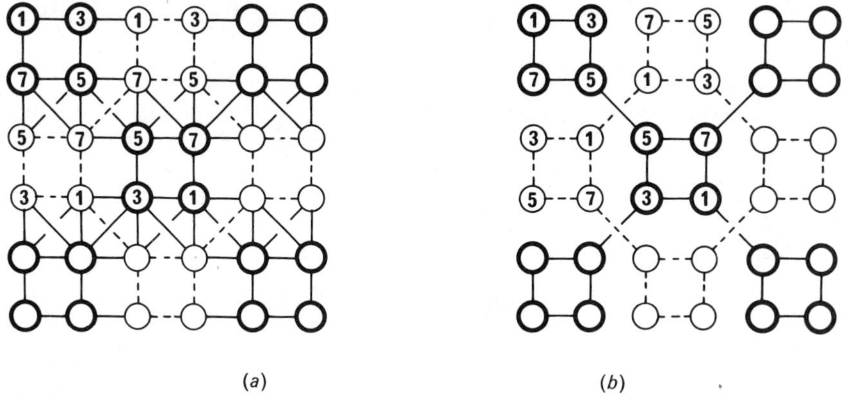

Fig. 11.1. Pairs of interpenetrating nets. (*a*) Two D-(10, 3)-*a* nets. (*b*) D and L forms of (10, 3)-*a* nets.

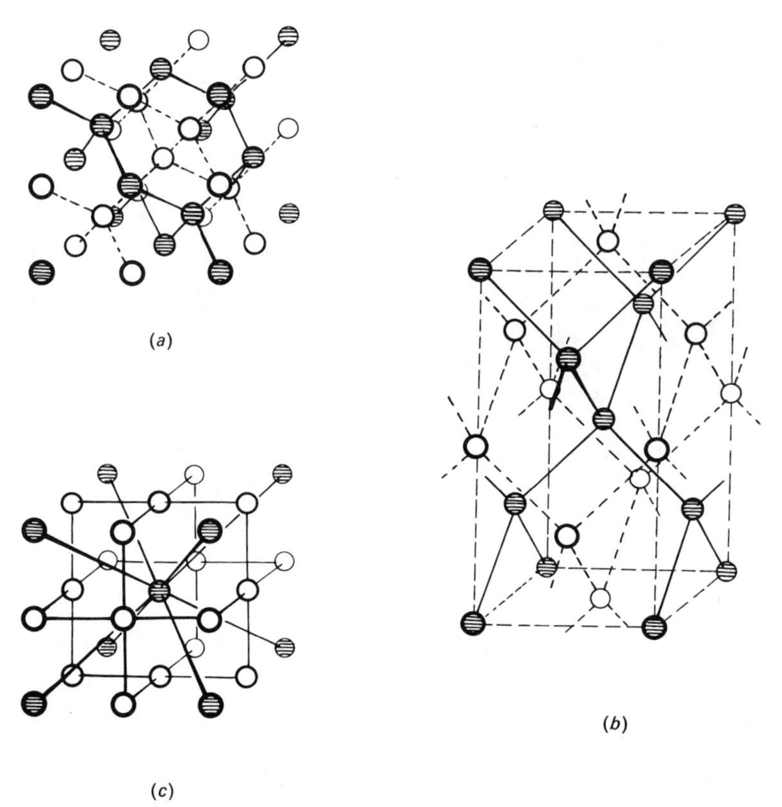

Fig. 11.2. (*a*) An *I* net as two interpenetrating diamond nets. (*b*) An *F* net as two interpenetrating 'extending-diamond' nets. (*c*) A *P* lattice as interpenetrating *I* and NbO nets.

two *different* nets, *I* and NbO (Fig. 11.2*c*). The component nets in Figs. 11.2*a* and *b* have been differentiated by preferentially linking a point to a limited number (only) of its equidistant nearest neighbors, 4 instead of 8 or 12, respectively; in Fig. 11.2*c* a 6-connected net has been broken down into a 4- and an 8-connected net. Since we are not pursuing this approach systematically, we include here some other examples.

In addition to the relation of Fig. 11.2*b* the cubic *F* (all-face-centered) net may be broken down into various numbers of interpenetrating nets that retain cubic symmetry. Taking an 8-fold cell, in which the coordinates of the points are those of the special position 32(*b*) in $Fd3c$, we may separate the 32

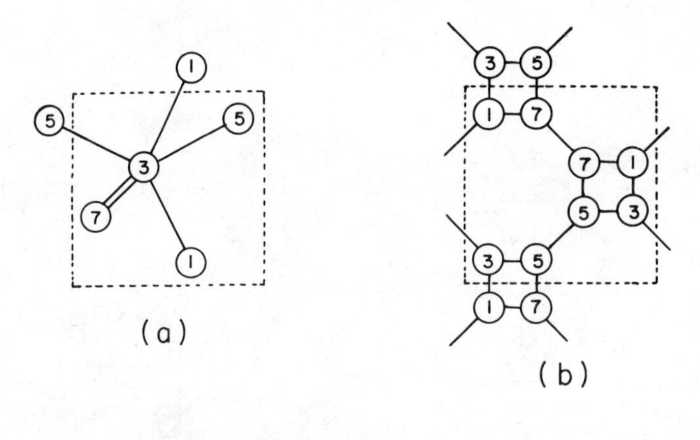

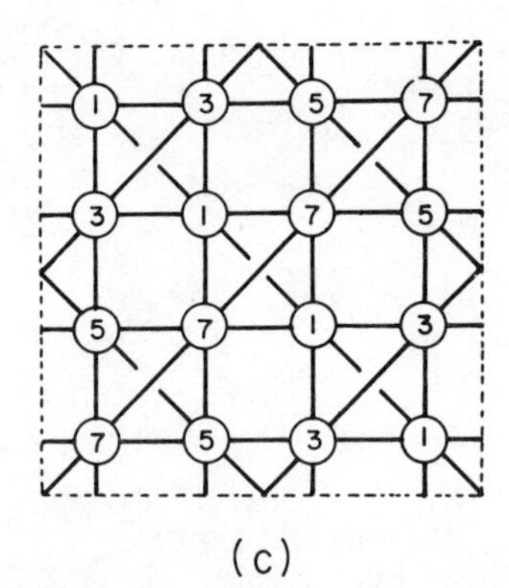

Fig. 11.3. Equivalent points in cubic space groups. (*a*) 4(*a*) in $P4_132$; (*b*) 8(*a*) in $I4_132$; (*c*) 16(*c*) in $F4_132$.

points into the following combinations:

eight 6-connected nets:	$4(a)$ in $P4_132$
four 3-connected nets:	$8(a)$ in $I4_132$
two 6-connected nets:	$16(c)$ in $F4_132$

Unit cells of the component nets are shown in Fig. 11.3. The 3-connected net is (10, 3)-*a*, and there are two ways of interpenetrating four such nets, the individual nets being either four D (or L) or two D and two L nets (Fig. 11.4). The two 6-connected nets that interpenetrate to form the equivalent positions

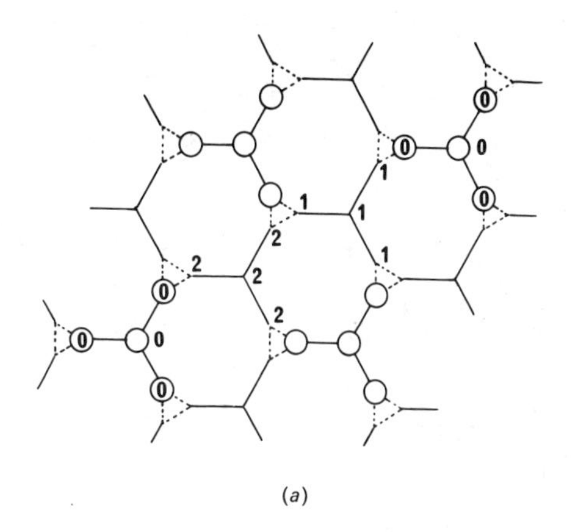

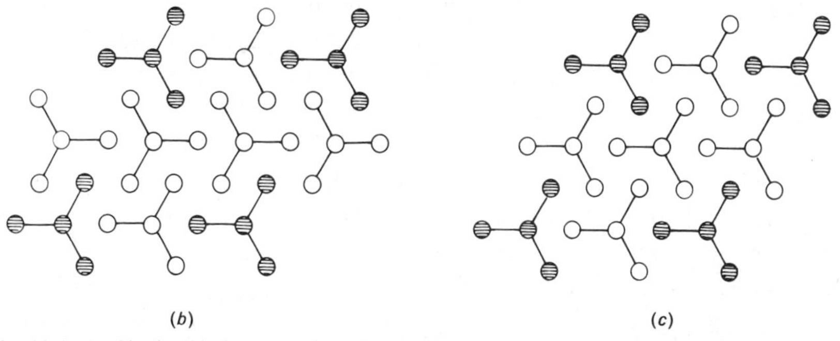

Fig. 11.4. (*a*) Single (10, 3)-*a* net viewed along 3_1 axis. Numerals indicate heights of points above the plane of the paper as multiples of $c/3$, where c is the repeat distance along the helix. (*b*) An *F* net as four interpenetrating (10, 3)-*a* nets, two D and two L. (*c*) An *F* net as four interpenetrating (10, 3)-*a* nets, all D (or L). The points in (*b*) and (*c*) are only those at one height (say, 0), the four groups of shaded circles belonging to one net.

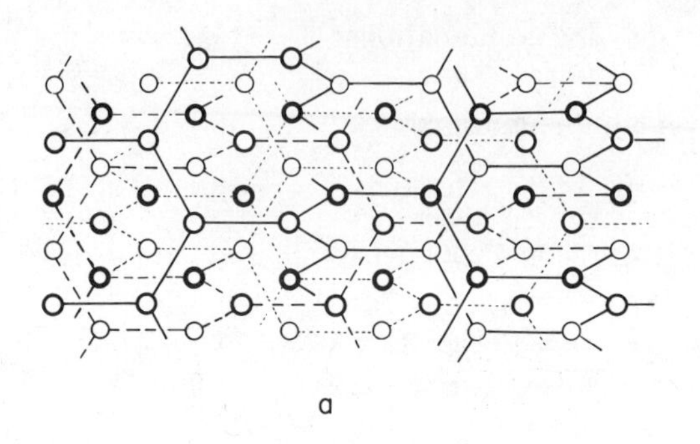

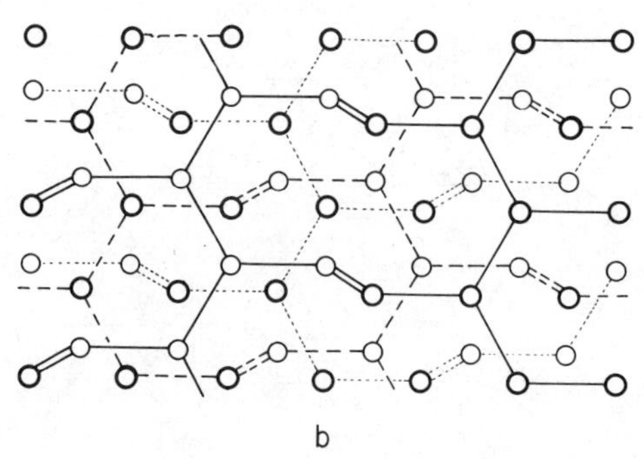

Fig. 11.5. Breakdown of (*a*) cubic and (*b*) hexagonal closest packing of equal spheres into three nets (10, 3)-*b*.

corresponding to the *F* net are basically of the diamond type. The equivalent points 16(*c*) in $F4_132$ are the midpoints of the links of the diamond net, and when joined as in Fig. 11.3*c* they outline tetrahedra around the points of that net, the edges of the tetrahedra forming the 6-connected net. Note also the breakdown of the *F* net into *three* nets (10, 3)-*b* (not the configuration with tetragonal symmetry but a "sheared" form with interaxial angle $109\frac{1}{2}°$ instead of 90°) and the breakdown of the h.c.p. net into *three* (10, 3)-*b* nets having a configuration with noncoplanar bonds. Projections of these two systems are shown in Fig. 11.5.

We now discuss two aspects of interpenetrating nets. (1) It is of interest to know whether the known structures consisting of two or more interpenetrating nets are special cases or only examples of a much larger class of possible structures. We need to set down the conditions to be satisfied if two or more (in the limit an indefinitely large number of) identical nets can interpenetrate. (2) In Part 10 the concept of different degrees of interpenetration of nets was mentioned, though it was not suggested how this feature of composite nets should be defined. This question is of interest in connection with 3D polyhedra and their complements, for the nets forming the bases of such pairs of space-filling polyhedra obviously illustrate the maximum degree of interpenetration of two nets. We describe such pairs of nets as *complementary* (*interpenetrating*) *nets*; they may be identical or different nets.

Interpenetration of nets of the same kind

The formation of two interpenetrating anticristobalite nets in Cu_2O is made possible by the large distance between atoms at the nodes of the (diamond) net (O–Cu–O, 3.7 Å) compared with normal bond lengths (e.g., C–C, 1.54 Å). This condition, together with the structure of β-quinol represented as a pair of identical interpenetrating 3-connected nets in which one-third of the bonds represent the long axes of molecules ($5\frac{1}{2}$ Å) and the remainder (in the hexagonal rings) O–H–O bonds (2.7 Å), and later the formation of six interpenetrating nets in [Cu(adiponitrile)$_2$]-NO_3, suggested the study of possible interpenetration of nets in which some (or all) bonds forming the basic net are (very) long compared with normal interatomic bond lengths. It was noted, for example, that if in the diamond net (referred to a b.-c. tetragonal cell) we join points separated by a translation of $a/2 + 3c/4$ instead of $a/2 + c/4$, the lines and points no longer represent a single net but three interpenetrating nets. This is illustrated in Fig. 11.6 for the net 4.14^2, which can be described as the tetragonal analogue of the trigonal 6.10^2 (β-quinol) net and is derived from the diamond net by replacing all points by squares with their planes parallel. Figures 11.6*a* and *b* show single nets, with their elevations at Figs. 11.6*c* and *d*. A structure of type *b*–*d* would seem possible for a long molecule with terminal OH, CO.NH, or other hydrogen-bonding groups, the squares representing closed circuits of hydrogen bonds analogous to the 6-rings in β-quinol.

It is evident that the number of interpenetrating nets in this type of structure is not restricted to three, for an indefinitely large number of the nets of Fig. 11.6*a* or *b* could interpenetrate. The first of these composite structures derived from Fig. 11.6*a* has two interpenetrating nets, and the crystallographic repeat distance along the vertical axis has dropped to $c/2$ (Fig. 11.6*e*). For n interpenetrating nets the repeat distance is c/n, and the unit cell (dotted rectangle in Fig. 11.6*f*) contains n ($1/n$th parts of molecules). A sketch of the

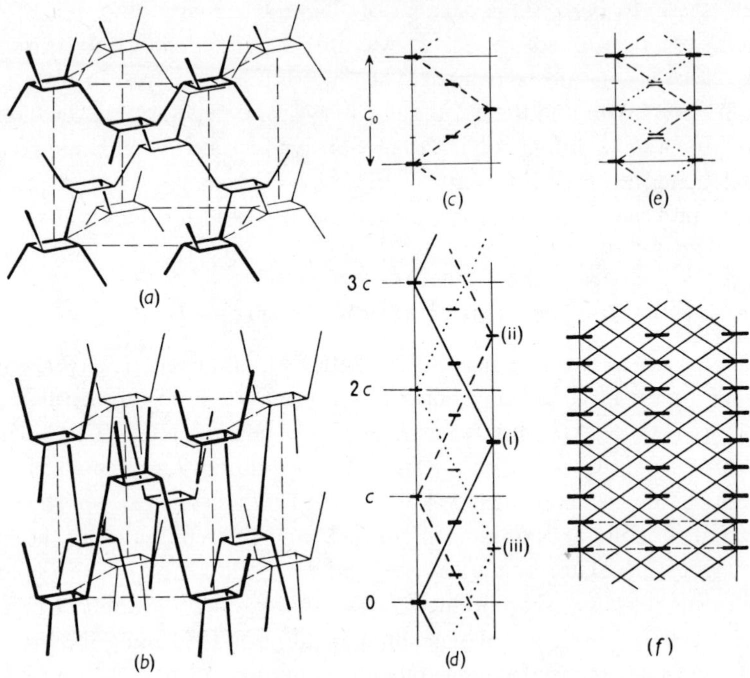

Fig. 11.6. Structures based on the 3-connected net 4.14^2 (p. 80).

type of structure so formed is given in Fig. 11.7*a*. Figure 11.7*b* shows the type of structure that would be formed by the interpenetration of large numbers of (10, 3)-*b* nets; other examples were given in Part 4.

To examine quite generally the possibility of the interpenetration of nets as a purely geometrical problem, we suppose the net to have its most symmetrical configuration, with all links of equal length and at each point the most symmetrical arrangement of links possible for a particular net. The conditions for interpenetration of an indefinite number of identical nets are as follows: (*a*) there must be a projection of the net in which no links intersect, (*b*) all links that superpose in the projection must be parallel in the 3D net, and (*c*) no links in the net are parallel to the direction of projection.

The projection direction consistent with these requirements corresponds to the direction of translation of the nets relative to one another. Condition *c* is obviously a limiting case; in the case of an actual crystal structure, metrical considerations would determine how nearly parallel to the projection direction the molecules could lie. To illustrate the application of these

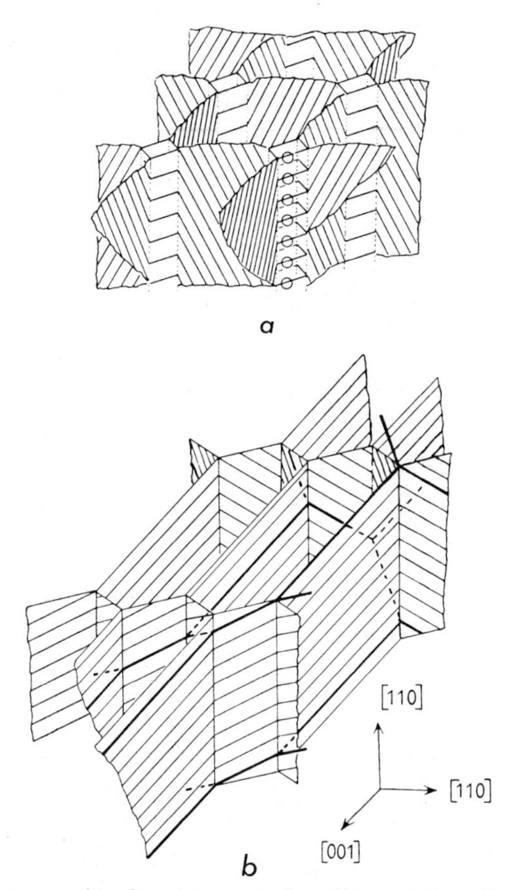

Fig. 11.7. (*a*) Structure resulting from interpenetration of large numbers of nets 4.14^2. (*b*) System of interpenetrating nets (10, 3)-*b*. The axes show the relation of this diagram to the tetragonal unit cell of a single net.

conditions we note that the projection of the *P* net along [100] is consistent with *a* and *b* but not with *c*, along [110] with *a* and *c* but not *b*, but along [111] with all three conditions. Projection of the diamond net along [110] does not satisfy *b*, but projection along [100] satisfies all conditions.

Since the projections are plane nets we may list the 3D nets which can interpenetrate in the way shown in Table 11.1. Two points can be noted here. A number of nets may project as the same plane net because a polygon in a plane net may represent a closed *n*-ring or a helical array of points in the 3D net, as shown in Fig. 11.8 for the three nets in Table 11.1 that project as the

Table 11.1 Plane nets corresponding to projections of single 3D nets or to systems of indefinite numbers of identical interpenetrating nets

Plane net*	3D net	Direction of projection
3^6	*P* lattice, (4, 6)	[111]
4^4	Diamond, (6, 4)	[100]
6^3	3-connected (12, 3)	[0001]
	3-connected (10, 3)-*b*	[110]
	3-connected 6.10^2	[0001]
3.9^2	3-connected (10, 3)-*a*	[111]
4.8^2	3-connected (10, 3)-*a*	[100]
	3-connected 4.14^2 (Fig. 6.9)	[001]
3.12^2	3-connected 3.20^2 (Fig. 6.2)	[111]
4.6.12	3-connected 4.12^2 (Fig. 6.3)	[111]
$3^2.6^2$	NbO	[111]
$3.4^2.6$	Fedorov (Fig. 9.20*a*)	[111]

* Symbols are point symbols except for uniform 3D nets (*n*, *p*).

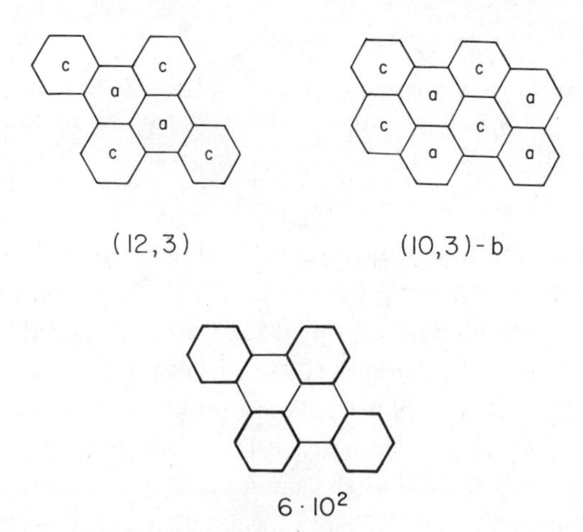

Fig. 11.8. Three 3-connected 3D nets that project as the plane 6^3 net. Heavy lines indicate closed 6-gons in the 3D net.

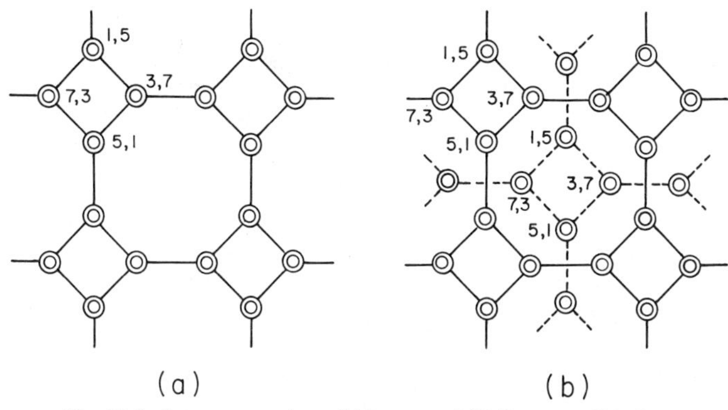

Fig. 11.9. Interpenetration of (*a*) two and (*b*) four nets (10, 3)-*a*.

plane 6-gon net. Conversely a given 3D net may project in different directions as different plane nets as, for example, (10, 3)-*a* as 3.9^2 along [111] and 4.8^2 along [100]. The latter case includes the interpenetration of *two* such nets related by a translation of $a/2$; it does not, however, include the interpenetration of *four* nets to give the *F* net. For whereas Fig. 11.1*a* can be redrawn as in Fig. 11.9*a* and is consistent with our conditions, the four interpenetrating nets related by translations of $a/2$, $b/2$, and $c/2$ would project in the [001] direction as in Fig. 11.9*b*, with links intersecting in the projection.

Examples of crystal structures illustrating some of the nets of Table 11.1 are given later.

We mention here a family of nets that are the simplest topological representations of the structures of the closely related fibrous zeolites edingtonite, thomsonite, and natrolite (see, e.g., A. F. Wells, *Structural Inorganic Chemistry*, 4th ed., Clarendon Press, Oxford, 1975, p. 828). The common structural feature is a chain formed from tetrahedral $(Si,Al)O_4$ groups that forms only a limited number of cross-connections (shared vertices) to other similar chains. These horizontal cross-links are in pairs separated by a distance $c/4$ along the chain, and the pairs along a given chain are in perpendicular planes. The simplest net corresponding to such structures therefore results from replacing the points of a *P* lattice by pairs of points forming four coplanar links as in Fig. 11.10*a*, giving the net of Fig. 11.10*b*. This net can be interpenetrated by a similar net (cf. 2 *P* nets → *I* net), the translation relating one net to the other being dependent on the value of the variable parameter z/c. The single net represents the structure of edingtonite, the vertical lines indicating the chains of tetrahedra; a pair of interpenetrating nets represents the structure of ice-VI.

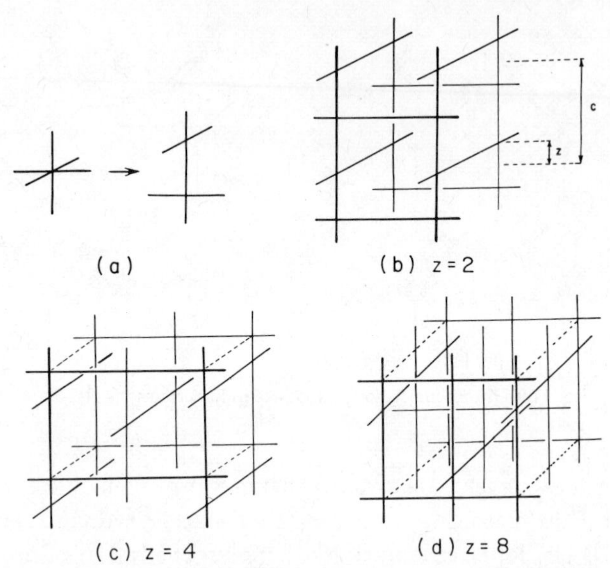

Fig. 11.10. Networks related to the structures of zeolites (see text).

The more complex nets of this family (Fig. 11.10*c* and *d*) represent in a similar way the structures of thomsonite and natrolite.

Complementary nets

It seems reasonable to suppose that two nets interpenetrate to the maximum extent when a link of one net passes through each shortest circuit of links of the other net. We assume that in each net all points are equivalent. The first nets to be studied are obviously those corresponding to Fedorov's five space-filling polyhedra, whose edges enclose similar, and similarly oriented, polyhedral compartments. The nets formed by the edges of space-filling rhombic dodecahedra and of space-filling elongated dodecahedra are $\left\{4, \begin{matrix}4\\8\end{matrix}\right\}$ and $\left\{\begin{matrix}4 & 4\\6, & 5\end{matrix}\right\}$ respectively; and having differently connected vertices, these nets are excluded by our requirement that all points be equivalent. We are left with the space-filling arrangements of cubes, hexagonal prisms, and truncated octahedra. However it is not necessary that the original net contain polyhedral cavities; we also have to study nets in which there are cavities surrounded by "pseudo-polyhedral" arrangements of rings. The first of these is the diamond net, in which there is a tetrahedral arrangement of four 6-rings around the points $(\frac{1}{2}\frac{1}{2}\frac{1}{2})$, and so on; its complement is an identical net displaced by $a/2$ relative to the first net. The next is the NbO net, with an "octahedral" arrangement of eight 6-rings around (000) and $(\frac{1}{2}\frac{1}{2}\frac{1}{2})$, the complement being the *I* net, as already shown in Fig. 11.2*c*. The five pairs of complementary nets are listed in Table 11.2. The two nets that are identical to their complements were presented in Table 11.1, for they are simply special cases of nets that satisfy conditions *a*, *b*, and *c* discussed in the previous section.

Table 11.2 Pairs of complementary nets

3D net	Complementary net	Notes
1* (4, 6)	(4, 6)	Edges of space-filling by cubes (*P* lattice)
2* (6, 4)	(6, 4)	Diamond net
3 $\left(\begin{matrix}6\\8\end{matrix}, 4\right)$	(4, 8)	NbO and *I* nets
4 $\left(\begin{matrix}4\\6\end{matrix}, 5\right)$ or $\left(\begin{matrix}4\\6\end{matrix}, 3+2\right)$	$\left(\begin{matrix}3\\4\end{matrix}, 8\right)$ or $\left(\begin{matrix}3\\4\end{matrix}, 6+2\right)$	Edges of space-fillings by hexagonal prisms and trigonal prisms
5 $\left(\begin{matrix}4\\6\end{matrix}, 4\right)$	(3, 14) or (3, 8 + 6)	Edges of space-filling by truncated octahedra

* Net and complement identical.

If the links of a net are the edges of a space-filling arrangement of polyhedra, the value of p must be 4 or more; but since we find complements of nets that do not contain true polyhedral cavities, it is necessary to determine whether there are complements to 3-connected nets. We shall examine in detail one 3-connected net, and of the two simplest we choose the net (10, 3)-*b* ($ThSi_2$) because it is relatively easy to visualize. The links of this net partially delineate pseudo-octahedral compartments of the type shown in Fig. 11.11*a*. Each of these compartments is surrounded by (and shares "faces" with) eight similar compartments whose centers are the points (000), $(\frac{1}{2}\frac{1}{2}\frac{1}{2})$, $(\frac{1}{2}0\frac{3}{4})$, and $(0\frac{1}{2}\frac{1}{4})$. These are the positions of the Th atoms in $ThSi_2$. When connected together, these points form an 8-connected net in which all links are of length a (Fig. 11.11*b*). It is now clear that the net that stands in a complementary

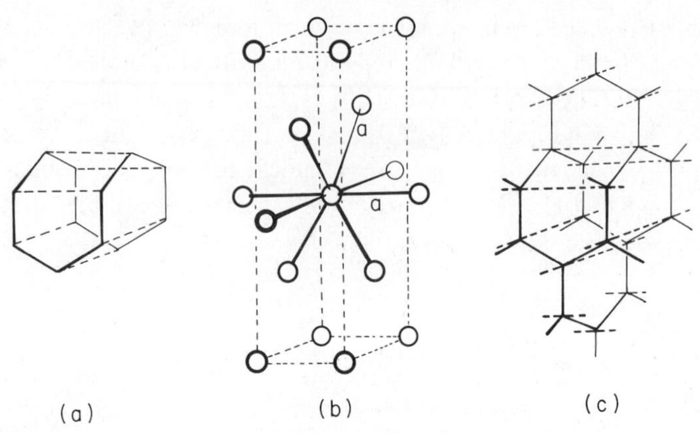

Fig. 11.11. The nonexistence of a net complementary to (10, 3)-*b* (see text).

relation to this 8-connected net is not the (10, 3)-*b* net but the 5-connected net formed by adding the links shown as broken lines in Fig. 11.11*a* and *c*. In this 5-connected net the links from each point are three coplanar (at 120°) of length $a/\sqrt{3}$ and two of length a that complete a trigonal bipyramidal arrangement. The links of this 5-connected net outline a space-filling arrangement of 3-connected octahedra ($f_4 = 4$, $f_5 = 4$), and those of the complementary 8-connected net outline a space-filling arrangement of regular trigonal prisms. These prisms are arranged in layers perpendicular to the tetragonal c axis, the 3-fold axes of the prisms in successive layers being parallel alternately to the tetragonal a and b axes. The close relation between this pair of complementary (3 + 2)- and 8-connected nets to pair 4 of Table 11.2, which are (3 + 2)- and (6 + 2)-connected, is more easily seen from the models in Figs. 11.12 and 11.13. The relation of the model of Fig. 11.12 to the structure of $ThSi_2$ is similar to that of the model of Fig. 11.13 to the AlB_2 structure.

In the cubic net (10, 3)-*a* a pseudo-polyhedral compartment can be found having six 6-gon and two 10-gon "faces," suggesting a complementary 8-connected net. As in the case of (10, 3)-*b*, the net of which an 8-connected net would be the complement is not the original 3-connected net but one with additional links, making the net 5-connected. We therefore conclude that the net (10, 3)-*a* has no complement in the sense defined earlier. However in this case there is no unique center to the pseudo-polyhedral compartment because the 4_1 (or 3 or 3_1) axes do not intersect; for this reason we dealt in more detail with the simpler case of the net (10, 3)-*b*.

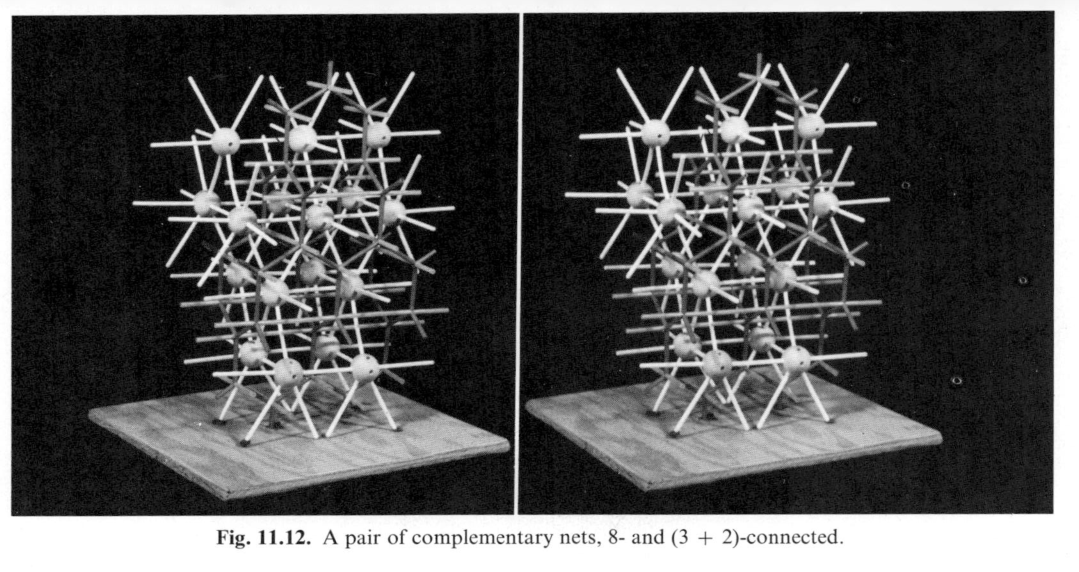

Fig. 11.12. A pair of complementary nets, 8- and (3 + 2)-connected.

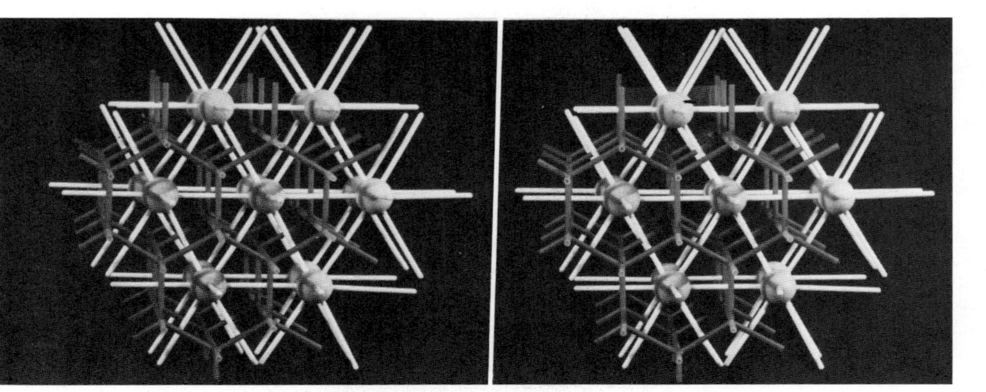

Fig. 11.13. A pair of complementary nets, (6 + 2)- and (3 + 2)-connected.

Crystal structures based on interpenetrating nets

In a number of crystals we can distinguish two or more interpenetrating 3D frameworks the atoms of which are not connected by bonds of the same kind as those within a framework. In the extreme case there is strong ionic–covalent bonding within the framework but only van der Waals bonding between atoms of different frameworks (as in cuprite). It is convenient to describe in the same way other structures in which the distinction

Table 11.3 Structures built of two or more interpenetrating 3D nets

I. Two or more nets of the same topological type
- *a.* Nets identical
  - (10, 3)-*a*, D and L nets forming 3D racemate: no example
  - (10, 3)-*b*, two nets—neptunite[1]
  - 6.10^2, two nets—β-quinol[2]
  - (3, 4)-connected nets, four nets—*P* phase,[3a] δ phase[3b]
  - Diamond or cristobalite, two nets—Cu_2O, ice-VII,[4] borates[5]
    six nets—$[Cu(adiponitrile)_2]NO_3$[6]
  - Edingtonite, two nets—ice-VI[7]
- *b.* Nets of different compositions
  - Diamond, $Hg_2Nb_2O_7$ (pyrochlore structure)[8]

II. Two complementary nets (of different topological types)
- *a.* NbO and *I* nets
- *b.* (3 + 2) and (6 + 2) nets

(*a* and *b*:) No examples known.

III. Combinations of I and II
NbO net and two interpenetrating diamondlike nets forming a pseudo-body-centered arrangement
$Bi_3GaSb_2O_{11}$[9]

REFERENCES

1. E. Cannillo et al., *Acta Crystallogr.*, 1966, **21**, 200.
2. H. M. Powell, *J. Chem. Soc.* (*London*), 1950, **298**, 300.
3. (a) D. P. Shoemaker et al., *Acta Crystallogr.*, 1957, **10**, 1; (b) C. C. Shoemaker and D. P. Shoemaker, *Acta Crystallogr.*, 1963, **16**, 997.
4. E. Whalley, D. W. Davidson, and J. B. R. Heath, *J. Chem. Phys.*, 1966, **45**, 3976.
5. J. Krogh-Moe, *Acta Crystallogr.*, 1965, **18**, 1088; 1968, **B24**, 179; 1969, **B25**, 2153.
6. Y. Kinoshita et al., *Bull. Chem. Soc. Japan*, 1959, **32**, 1221.
7. B. Kamb and B. L. Davis, *Proc. Nat. Acad. Sci.* (*U.S.*), 1964, **52**, 1433.
8. A. W. Sleight, *Inorg. Chem.*, 1968, **7**, 1704.
9. A. W. Sleight and R. J. Bouchard, *Inorg. Chem.*, 1973, **12**, 2314.

between the intra- and interframework bonds is less clear-cut, however, if only as a help in describing the basic topological features of the structures. Since the number of structures known to be of this type is not very large, an elaborate scheme of classification is not justified. In Table 11.3 the structures are grouped according to the nature of the component frameworks (nets). The following subsections give additional information about the structures; more details can be found in the literature references given in the table.

Table 11.3; Class Ia

Only two examples of structures consisting of interpenetrating 3-connected nets are noted. If one-third of the links in the net (10, 3)-*b* are much longer than the others, the net projects along [111] as shown in Fig. 11.14. The mineral neptunite has the composition $M^I_4M^{II}_2(TiO)_2Si_8O_{22}$, where M^I represents an alkali metal and M^{II} a doubly charged ion of Fe, Mg, or Mn. The structure consists of two interpenetrating nets formed from SiO_4 tetrahedra, one-half of which share three vertices [and are situated at the nodes of a (10, 3)-*b* net]; the remainder share two vertices and are situated in pairs along the longer links of Fig. 11.14. Each net has the composition Si_8O_{22} and the remaining O atoms, together with O atoms of the frameworks, complete octahedral coordination groups around the Ti ions which, like the M^I and M^{II} ions, do not form part of the frameworks.

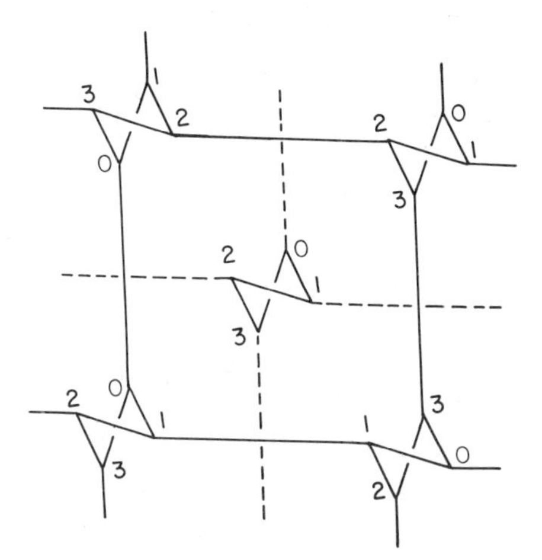

Fig. 11.14. Pair of interpenetrating nets forming the basis of the structure of neptunite.

The crystal structures of certain dihydroxy compounds may be described as 3-connected nets in which two-thirds of the links represent hydrogen bonds (Part 3).

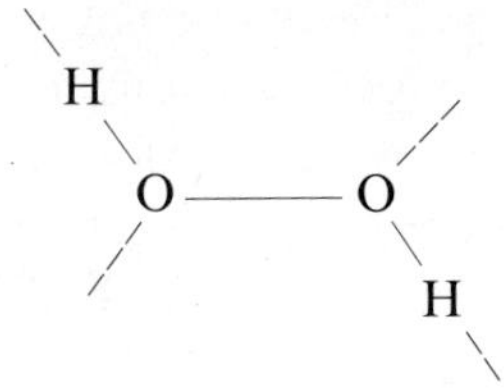

In the β-quinol clathrates two 6.10^2 nets (Fig. 6.13), in which the hexagons represent rings of O–H···O bonds, interpenetrate; the cavities so formed are occupied by foreign molecules such as SO_2 or CH_3OH or by atoms of argon.

Certain intermetallic compounds formed by transition metals can be described in terms of "major ligand" nets that form systems of four interpenetrating (3, 4)-connected nets in the δ phase (Mo–Ni) and *P* phases.

A number of structures consist of or contain frameworks based on the diamond net or the cristobalite net, which has the composition AX_2 and results from placing X atoms along the links of the diamond net. In Cu_2O and ice-VII, oxygen atoms occupy the nodes of the diamond net and the Cu or H atoms are situated along the links, the structures consisting of two identical interpenetrating nets. In a number of boron-rich borates boron-oxygen complexes such as (*a*) or (*b*) at each node of a diamondlike net are

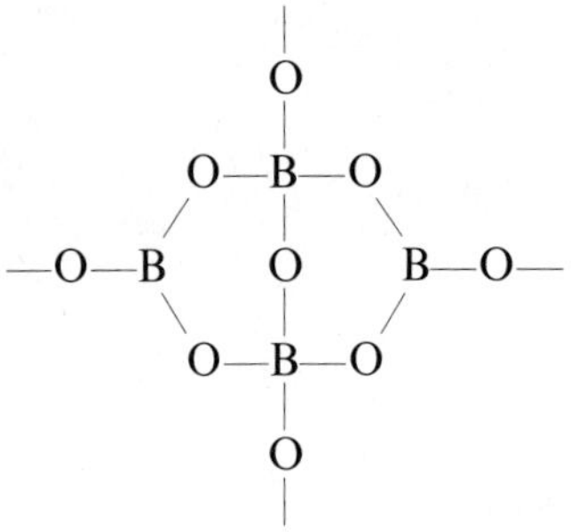

(*a*)

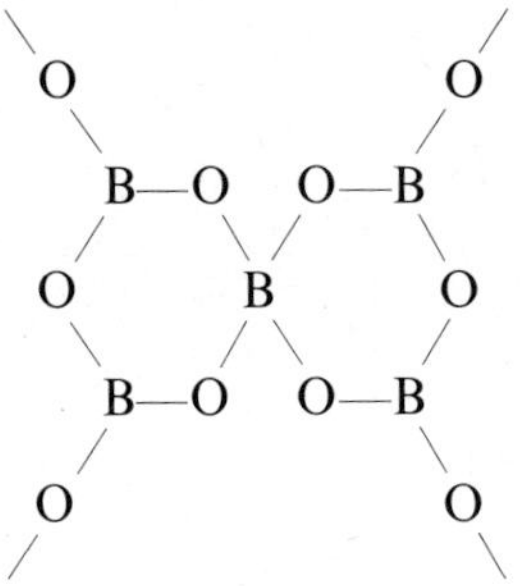

(*b*)

joined to four other units via the extracyclic O atoms, as in $Li_2B_4O_7$ (from units *a*) and KB_5O_8 (from units *b*). Units of two kinds alternate in $Ag_2B_8O_{13}$. In these crystals the anions are systems of two interpenetrating 3D framework ions.

The structure of $[Cu(adiponitrile)_2]NO_3$ is notable for being built of six identical interpenetrating nets in which Cu atoms are at the nodes of diamondlike nets and are linked through $-NC(CH_2)_4CN-$ ligands.

In the fibrous zeolite edingtonite, $Ba(Al_2Si_3O_{10})\cdot 4H_2O$, there is a (single) framework of linked (Si, Al)O_4 tetrahedra as described on page 159. Ice-VI consists of two interpenetrating identical frameworks (built from hydrogen-bonded water molecules) of the same kind as in edingtonite.

Table 11.3; class Ib

The pyrochlore structure can be described in terms of a 3D framework BX_3 formed from vertex-sharing BX_6 octahedra arranged along the links of the diamond net. The seventh X atom in the formula $A_2B_2X_7$ does not form part of this framework. In $Hg_2Nb_2O_7$ the seventh O atom and the Hg atoms form a cuprite (anticristobalite)-like framework that interpenetrates the octahedral framework, as indicated by writing $(Hg_2O)(Nb_2O_6)$.

Table 11.3, classes II and III

No examples are known of structures in class II, but our sole example in class III is closely related to the Cu_2O structure in I*a* and to class II*a*. If two anticristobalite A_2X structures interpenetrate symmetrically as in Cu_2O, there is formed a pseudo-body-centered structure in which each X has eight equidistant neighbors but is bonded to only four of them (through Cu atoms in Cu_2O or H atoms in ice-VII). A third net, of the NbO type, can interpenetrate this system of two interpenetrating diamondlike nets. In $Bi_3GaSb_2O_{11}$ the former is an octahedral framework AX_3, whereas the latter have a complex structure built from tetrahedral Bi_4 groups joined through O and Bi atoms, with the composition $Bi_3O_2(Bi_{3/2}O)$. The structural formula therefore can be written $(Bi_{3/2}O)_2(GaSb_2O_9)$.

12

Relation of three-dimensional nets to "open sphere packings"

The most open (least dense) packings of equal, equivalent spheres for 3- and 4-coordination were studied by Heesch and Laves,[1] who required all the spheres to be crystallographically equivalent and to form one connected, infinite, periodic 3D system. Their condition for stability also implies that when a sphere is in contact with no more than three others, the four sphere-centers must be coplanar. Examples of even less dense packings of *non-equivalent* spheres were given by Melmore.[2,3] The nets formed by connecting the centers of adjacent spheres are examples of 3- and 4-connected nets, of which some are uniform nets. Heesch and Laves were interested in the *most* open packings of equivalent spheres, and they considered only the most symmetrical structures, with cubic, hexagonal, or rhombohedral symmetry. The open packings were called

3-coordination types: 3_1 and 3_2

4-coordination types: 4_1, 4_2, 4_3, and 4_4

The symbols refer to the topological type, that is, to the kinds of polygon that meet at each point in the underlying net, and there may be more than one packing corresponding to a particular symbol. The density of a sphere packing given in Table 12.1 is the fraction of the whole of space that is occupied by spheres.

Table 12.1 Open packings of equal, equivalent spheres

Net or H and L symbol	Point symbol	Symmetry	Density
(10, 3)-*c*	10^3	Hexagonal	0.269
(10, 3)-*b*	10^3	Tetragonal	0.233
(10, 3)-*a* } H and L 3_1	10^3	Cubic	0.185
(8, 3)-*a* } H and L 3_1	8^3	Hexagonal }	0.172
(8, 3)-*b* } H and L 3_1	8^3	Rhombohedral }	0.172
Fig. 6.4	6.8^2	Cubic	0.165
Fig. 6.3	4.12^2	Cubic	0.117
H and L 3_2	3.20^2	Cubic	0.056
H and L 4_1	{ 6^6	Cubic }	0.338
	{ 6^6	Hexagonal }	
H and L 4_2	{ $3.4.6^2.8^2$	Cubic	0.215
	{ 3.6^5	Hexagonal	0.244
H and L 4_3	$3^2.10^4$	Cubic	0.180
H and L 4_4	$3^3.12^3$	Cubic	0.123

Type 3_1. The cubic form is the regular configuration of the 3-connected net (10, 3)-*a*. The hexagonal and rhombohedral "variants" are the nets (8, 3)-*a* and (8, 3)-*b*.

Type 3_2. This arises from the cubic configuration of 3_1 by replacing each sphere by an equilateral triangular group of three spheres; it corresponds to the net 3.20^2.

Heesch and Laves' packing 3_2 is apparently the least dense packing of equal, equivalent spheres (on which point see Melmore, Ref. 3, p. 484), but at least two other packings of crystallographically equivalent spheres with densities intermediate between those of their packings 3_1 and 3_2 satisfy their condition that all three neighbors be coplanar with a given sphere. These two cubic nets, which are included in Table 12.1, have already been described under Archimedean 3-connected nets. Their relation to the packings 3_1 and 3_2 is seen from the topological diagrams of Fig. 12.1, which show the inter-bond angles in the underlying nets.

We also include in Table 12.1 the two other simplest uniform (10, 3) nets, the densities of which are greater than those of the forms of 3_1. The density

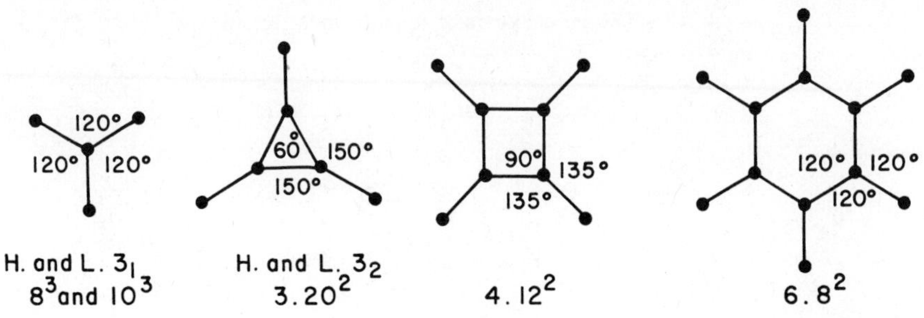

Fig. 12.1. Topological diagrams for 3-connected nets related to open sphere packings (3-coordination).

of 3_1 (cubic) is equal to one-quarter of that of closest packing (0.7405), for c.c.p. can be broken down into four cubic (10, 3)-*a* nets as described in Chapter 11.

Type 4_1. The cubic and hexagonal forms correspond to the cubic and hexagonal diamond structures, or the sphalerite and wurtzite structures for compounds AX.

Type 4_2. The cubic packing corresponds to the edges of the polyhedra in the space-filling by truncated tetrahedra, truncated cubes, and truncated cuboctahedra (Fig. 12.2). The hexagonal packing is illustrated as the 4-connected net 3.6^5 in Fig. 9.16.

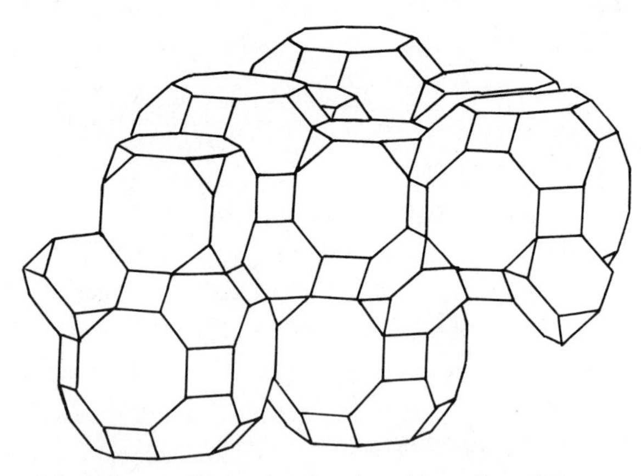

Fig. 12.2. The polyhedral space-filling related to the cubic variant of the open sphere packing 4_2 of Heesch and Laves.

Type 4_3. The cubic net corresponding to this packing arises by joining the midpoints of adjacent links in the cubic configuration of the net (10, 3)-*a*.

Type 4_4. This packing results from replacing the single spheres in the diamond packing 4_1 by tetrahedral groups of spheres. Further details (space groups and equivalent positions) of the cubic packings 4_1–4_4 are included in Table 10.1.

In Table 12.1 we give the density (0.233) of the sphere packing in which the centers of the spheres occupy the points of the tetragonal configuration of the net (10, 3)-*b*. On page 154 we showed configurations of this net corresponding to one-third of the positions of hexagonal or cubic closest packings of spheres, therefore a density of 0.247. By rotations about the links parallel to the tetragonal *c* axis, this net "shears" to more compact configurations, and the density of the sphere packing corresponding to the most compact configuration is 0.421.

References

1. H. Heesch and F. Laves, *Z. Kristallogr.*, 1933, **85**, 443.
2. S. Melmore, *Nature*, 1942, **149**, 412, 669.
3. S. Melmore, *Mineralog. Mag.*, 1949, **28**, 479.

13

Review of three-dimensional nets

Since the general remarks on nets made in the introductory sections in some respects anticipated conclusions reached in later sections, it is appropriate to comment here on certain aspects of nets in the light of the results of the systematic studies we have described. Our observations concern three topics.

A general classification of three-dimensional nets

In Part 7 we suggested that all periodic, uniform 3D nets may be placed on a diagram in which we plot along three axes n, p, and either x or y; Fig. 13.1 is similar to Fig. 3 of Part 7 but includes points representing nets discovered after 1963, when that paper was written. Since all the links in some nets do not have the same y value, and since we wish to represent a net by a single point on the diagram, we plot y_{mean}, which for some nets has a nonintegral value. In the base of Fig. 13.1 are found the regular solids, the regular plane nets, and the 3D nets having y equal to 2; nets with higher values of y lie at higher levels. We have seen that the x and y values referring to a particular n-gon (^nx and ^ny) represent only the first stage in the topological description of a net. They do not in all cases distinguish one uniform (n, p) net from others. For example, on the vertical line corresponding to (10, 3) nets, each of the points $y = 3\frac{1}{3}$, $y = 8$, and $y = 10$ represents one net only [the nets (10, 3)-*c*, *g*, and *a*, respectively), whereas the point $y = 6\frac{2}{3}$ represents all the nets (10, 3)-*b*, *d*, *e*, and *f*. Figure 13.1 does not include points for all the nets described earlier.

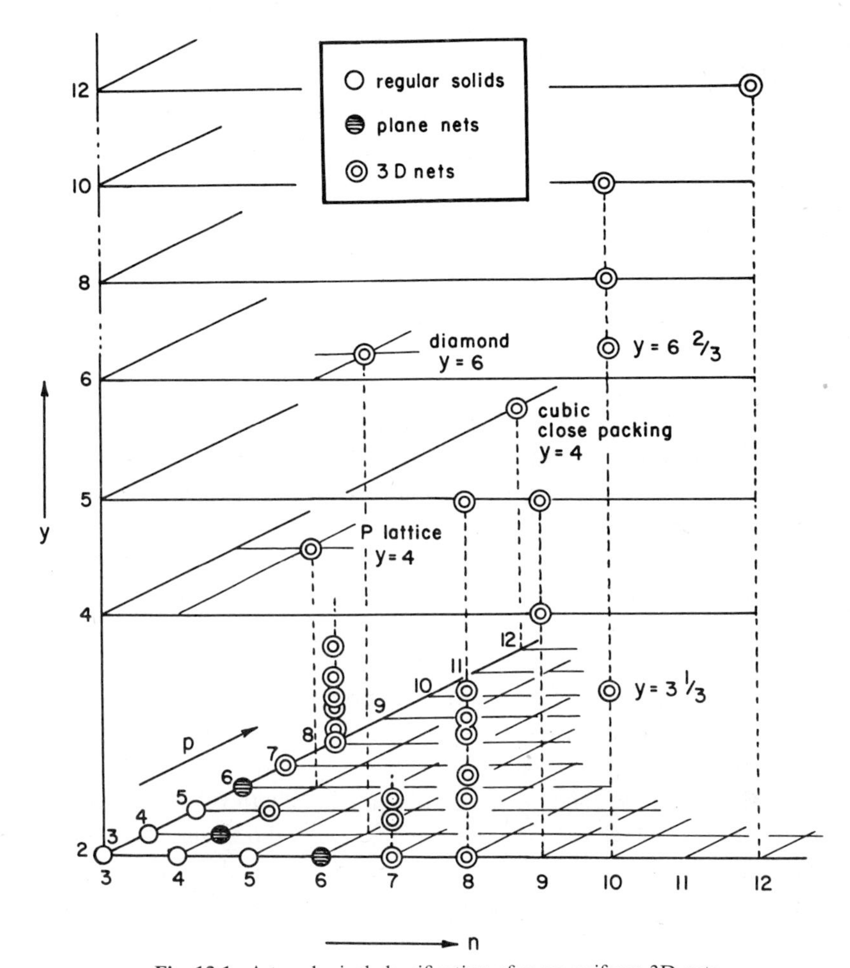

Fig. 13.1. A topological classification of some uniform 3D nets.

The symmetries of nets

From a crystal-chemical standpoint it is justifiable to compare quite generally all nets of the same connectedness. For example, the structure of a compound A_2X_3 in which each A atom is to be bonded to three X and each X to two A could be based on any 3-connected net (with X atoms along the links), and similarly for a compound AX_2 and 4-connected nets. However, the factors determining the choice of a particular net include the packing of the X atoms and the values of the X–A–X and A–X–A bond angles, to mention only the more obvious ones, and these are all neglected when describing the basic topology of the structure in terms of a 3-connected net. There is accordingly no direct connection between the topology of the net and the structure of a crystal. For example, we do not understand why the structures of certain compounds are based on the net (10, 3)-*a* and others on the net (10, 3)-*b* while that of boron trioxide is based on (10, 3)-*c*; all these nets are 3D systems of 10-gons. Certainly we do find a preference for simple nets as the bases of crystal structures, and in their most symmetrical configurations these do have high symmetry. Thus most of the (still rather few) crystal structures based on 3-connected nets are in fact based on the two simplest nets, that is, those with the smallest value of Z_t; and these nets have high topological symmetry and also high crystallographic symmetry in their most symmetrical configurations. Moreover, in spite of the neglect of all metrical factors that is implicit in describing a structure in terms of its basic topology, we find a close correspondence between the symmetry of the most symmetrical configuration of a net and that of the structures based on it. This correspondence between net and crystal structure also extends to a general similarity in cell dimensions. Each of the two simplest 3D 3-connected nets, (10, 3)-*a* and (10, 3)-*b*, has a second special configuration with interbond angles of 90° instead of 120°, the value for the most symmetrical configurations. We shall see shortly that we can tell from the symmetry and cell dimensions of a structure based on one of these nets which of the two special configurations forms the basis of the structure.

It is noteworthy that the uniform 3-connected net with the highest x value is the (12, 3) net, which is not yet known to form the basis of any actual crystal structure. Similarly, much use is made in crystal structures of the simplest (and most symmetrical) 3D 4-connected net, the diamond net.

Regarding the symmetry of uniform nets, we must distinguish between (*a*) nets in which there is equivalence of points and/or equivalence of links, and (*b*) nets that have a configuration in which all links are equal in length and all bond angles are those corresponding to the most symmetrical arrangement of bonds at each point. Nets of type *b* might be called "regular nets." These are quite different criteria. In the case of 3-connected nets *a* is a much more stringent condition than *b*, but the purely geometrical requirement *b* is also precise and is of interest when we survey nets as a group.

Nets of type a

As already indicated we can approach this first as a purely topological problem, that is, considering the net simply as a periodic 3D system of

Table 13.1 Some properties of uniform 3D 3-connected nets

Net	$^n x$ the same for all points	All points crystallographically equivalent in most symmetrical configuration	$^n y$ the same for all links	Net has a configuration with all links equal and all angles 120°
(7, 3)	*c*	—	*c*	—
(8, 3)	*a b c* i j k n*	*a b*	*c* i k n*	*a b c n*
(9, 3)	*a b*	—	*a b*	*a b*
(10, 3)	*a b c d f g*	*a b c d*	*a g*	*a b c*
(12, 3)	*a*	*a*	—	—

* This net has points with different values of ^{18}x and links with different values of ^{18}y.

connected points, without reference to the geometry (bond lengths and interbond angles). Consider first the points. We can list the nets that have the same value of $^n x$ for all points, separating thereby 17 of the 30 known uniform 3-connected nets (column 1 of Table 13.1). Alternatively we could list those nets in the most symmetrical configurations of which all the points are crystallographically equivalent; these number seven (column 2). Presumably the difference between columns 1 and 2 occurs because we have considered only $^n x$; we would expect the topological nonequivalence of the points to become evident if the x values were determined for other (larger) circuits. So far this point has been verified only for the net (8, 3)-*c*, which has different values of ^{18}x and ^{18}y for points and links that have the same values of $^8 x$ and $^8 y$, respectively. A thorough study of this aspect of all the nets appearing in column 1 but not in column 2 would be somewhat laborious and has not yet been made.

As in the case of the points, the topological study of the links has proceeded only far enough to take account of the n-gons. Column 3 lists the nets in which $^n y$ has the same value for all links. The crystallographic equivalence of the links is a matter of point symmetry. If all the links in a 3-connected net are to be equivalent, each point must lie on a 3-fold symmetry axis; note that this implies equal lengths of all links and equal angles between the links meeting at each point, but bond angles of 120° are not implied. The only net satisfying the last criterion is (10, 3)-*a*, which occupies a unique position among 3D 3-connected nets as the only one having a configuration in which all points are equivalent and all links are equivalent. This is the only net common to columns 2 and 3. A 4-connected net has the following condition for equivalence of all links: the points must lie at positions of appropriate 4-fold point symmetry (4/*mmm* for the points forming four coplanar bonds in the NbO net or 43*m* for the regular tetrahedral bonding in cubic diamond). The only *uniform* 4-connected net satisfying this criterion is the cubic diamond net.

Nets of type b

Column 4 of Table 13.1 lists the uniform 3-connected nets known to have a configuration in which all links are equal in length and all interbond angles are 120°. These nine nets include four with crystallographically nonequivalent points, and all except (10, 3)-*a* have crystallographically nonequivalent links.

The special forms of the nets (10, 3)-*a* and (10, 3)-*b* with mutually perpendicular bonds meeting at each point are of considerable interest in connection with the crystal chemistry of certain groups of compounds. These configurations of the nets can be derived from a portion of the primitive cubic lattice containing eight points by removing half the links. This operation may be carried out in many ways, to leave discrete cubes, chains, layers, or 3D nets. Of the numerous possibilities, the two shown in Fig. 13.2 correspond to the nets (10, 3)-*a* and *b*. The importance of these special 90° configurations, as opposed to the 120° configurations, lies in the possibility of recognizing them in structures based on these nets. We consider these two nets in turn.

Both configurations of the first net, which we call (10, 3)-*a* (120°) and (10, 3)-*a* (90°), have cubic symmetry. The former has been illustrated in Figs. 5.2 and 5.11; the space group is $I4_1 32$. A feature of the net is the presence of sets of 4_1 screw axes, and one set is evident in the projection of Fig. 5.2*b*.

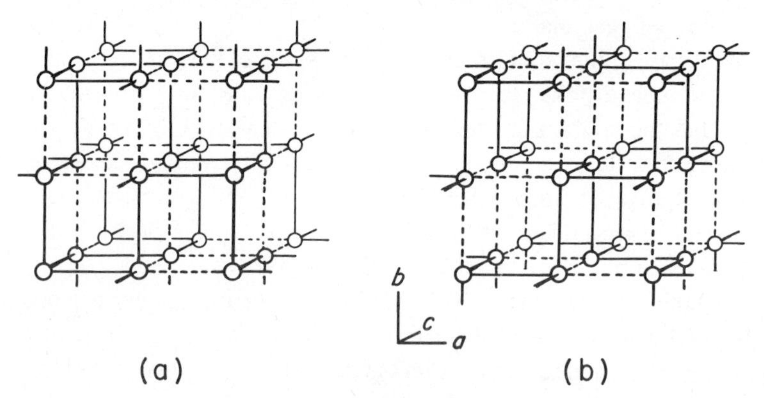

Fig. 13.2. The configurations of the nets (10, 3)-*a* and (10, 3)-*b* with 90° bond angles.

These screw axes are present in $SrSi_2$ and $CsBe_2F_5$; both compounds have eight 3-connected points per cell (Si and Be, respectively), and both have the same space group, $P4_132$. Presumably $P4_132$ is adopted instead of $I4_132$ because there is more freedom of arrangement of the helices in the former. In $I4_132$ the helices are not only similarly oriented (as they would be in a structure containing four points per cell) but they are related by a fixed translation of $(\frac{1}{2}\frac{1}{2}\frac{1}{2})$. Alternatively we may say that since the number of Sr atoms in $SrSi_2$ or of Cs in $CsBe_2F_5$ is equal to half the number of atoms (Si or Be) forming the 3-connected net, occupation of the 8-fold position 8(*a*) by the latter would require a 4-fold position for the former atoms, and there is no such position in $I4_132$. (A similar problem does not arise for $ThSi_2$ in $I4_1/amd$, for this space group has the required 4-fold position.)

In crystalline H_2O_2 the 4_1 helices are formed by the hydrogen bonds, which here constitute two-thirds of the links of the basic net. Since their length (2.78 Å) is so much greater than that of the remaining one-third of the links (which are the intramolecular O–O bonds, 1.47 Å long), the net is very distorted from the ideal configuration. The 4_1 axes are retained but the symmetry has dropped to tetragonal (a = 4.06 Å, c = 8.00 Å) and the space group is $P4_12_1$. The distortion of the net in this structure is so great that it is virtually 5-connected, for each O atom has also two other O atom neighbors at 2.90 Å.

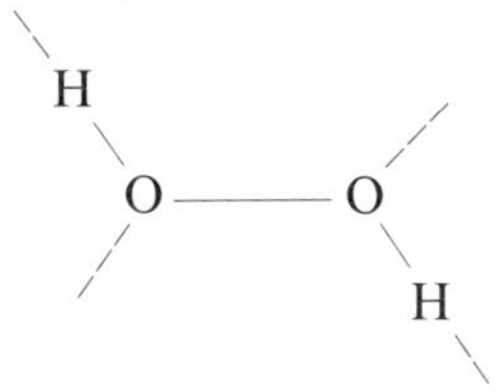

H
O——O
H

To assign a space group to the 90° configuration of (10, 3)-*a* shown in Fig. 13.2*a*, it is necessary to indicate the positions of the links, since the points themselves form a simple cubic lattice in all diagrams of this type. This is done by placing (2-connected) points at the midpoints of the links, which would be the positions of X atoms in a compound A_2X_3 based on this net or of the shared X atoms in a framework of composition A_2X_5 formed from tetrahedral AX_4 groups sharing three vertices. Such points are the set of equivalent positions 12(*b*) of $I2_13$ for the special value $x = \frac{1}{2}$. It is satisfactory to find that the structure of α-$Hg_3S_2Cl_2$ is based on the net (10, 3)-*a* (90°) and that the S atoms occupy 8(*a*) and the Hg atoms 12(*b*) in the space group $I2_13$.

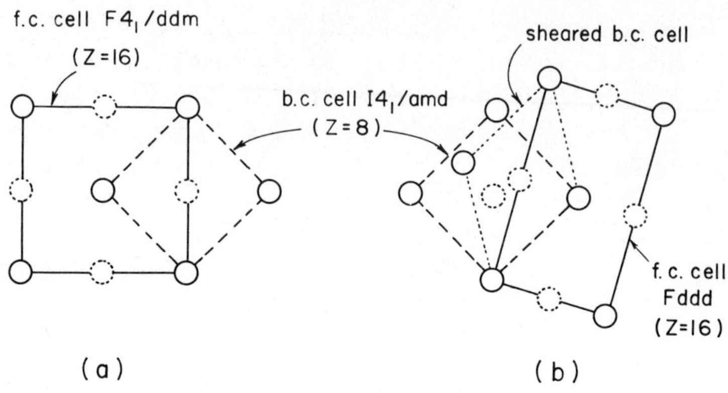

Fig. 13.3. (*a*) Relation between *F* and *I* cells of (10, 3)-*b* projected along the *c* axis. Dotted circles represent points *c*/2 above (below) the plane of the paper. (*b*) Relation between the b.c. cell of "sheared" configuration of (10, 3)-*b* (dotted lines) and the all-face-centered orthorhombic cell (full lines).

The net (10, 3)-*b* is somewhat more complex because the 120° form of this net can be "sheared," as described on page 154, a process involving rotations about the links that are parallel to the tetragonal *c* axis. The fully extended tetragonal form of the net represents the arrangement of Si atoms in α-$ThSi_2$. In sheared forms the 8-point b.c. tetragonal cell becomes nonorthogonal, but the sheared net may be referred to an all-face-centered orthogonal cell containing 16 points. These are alternative cells for the space group *Fddd* (Fig. 13.3). This relation is similar to that between the f.c. cubic and b.c. tetragonal cells of diamond. We may refer cubic diamond to a b.c. tetragonal cell containing four atoms in 4(*a*) of $I4_1/amd$, and conversely we may refer the net (10, 3)-*b* to a 16-point f.c. cell with space group $F4_1/ddm$, which is the alternative setting of the same space group. In both cases the most symmetrical orthorhombic subgroup is *Fddd*.

In fact we find that one form of P_2O_5 and also $(Zn_2Cl_5)(H_5O_2)$ both have the space group *Fdd*2, which is also that of GeS_2,* whose structure is also based on this net in the way described on page 128. The first two structures have 16 3-connected points (P or Zn atoms, respectively) in the unit cell, but GeS_2 has $Z = 24$ because here the Ge atoms are situated at the midpoints of the *links* of the original 3-connected net. The relation of these structures

* It has recently been shown (G. Dittmar and H. Schäfer, *Acta Crystallogr.*, 1976, **B32**, 1188) that low-GeS_2 is actually monoclinic (space group *Pc*). The structure is essentially the same as that of Zachariasen (ref. 10, Table 13.2).

to the 120° configuration of (10, 3)-*b* is also seen from the cell dimensions of these crystals. For the tetragonal configuration of the net $c/a = 2\sqrt{3} = 3.46$, and $\beta = 90°$. The corresponding figures for the b.c. nonorthogonal ($Z = 8$) cells of the other compounds are as follows:

	$a\,(=b)$	c	β (approx.)	c/a
P_2O_5	4.85 Å	16.3 Å	114°	3.36
$(Zn_2Cl_5)(H_5O_2)$	6.43 Å	22.90 Å	92°	3.56
GeS_2	6.77 Å	22.34 Å	119°	3.30

Note that the dimensions of the orthorhombic cells consistent with the setting *Fdd*2 (i.e., 2-fold axes parallel to *c* axis) are as follows:

	a	b	c
P_2O_5	16.3 Å	8.14 Å	5.26 Å
$(Zn_2Cl_5)(H_5O_2)$	9.26 Å	22.90 Å	8.91 Å
GeS_2	11.66 Å	22.34 Å	6.86 Å

The longest cell dimension in each case corresponds to the tetragonal *c* axis of (10, 3)-*b*. The reason for the adoption of the less symmetrical space group *Fdd*2 instead of *Fddd* cannot be given at this time. To accommodate the required numbers of atoms in the unit cells (e.g., 40 O, 40 Cl, or 24 Ge), it is necessary to occupy 8-fold in addition to 16- and/or 32-fold positions, and the numbers of variable parameters x, y, and z (e.g., one in the 8-fold position of *Fdd*2 but none in those of *Fddd*) show that there is much more freedom of movement in the lower space group. Studies of the packing of the larger ions (atoms) that are consistent with the required modes of linking of the structural units may throw some light on this subject.

On the other hand the unit cell of the ideal configuration (10, 3)-*b* (90°), Fig. 13.2*b*, is a cube, but the symmetry is of course not cubic. By placing points at the midpoints of the links, we determine the space group as $C2/c$, which is the space group of $La_2Be_2O_5$, whose structure is based on this configuration of the net.

A similar correspondence, closer in some cases than in others, is seen between the symmetries of the most symmetrical configurations of other nets and those of the structures based on them. The known examples are summarized in Table 13.2. This is not the appropriate place for a more detailed discussion of these structures, and a review of the crystal chemistry of structures based on 3-connected nets has been published. (*Acta Crystallogr.*, 1976, **B32**, 2619).

Table 13.2 Structures based on 3D 3-connected nets

	Most symmetrical configuration					
Net	Space group	Position	c/a	Compound	Space group	Reference
(10, 3)-*a* (120°)	$I4_132$	8(*a*)		$SrSi_2$	$P4_132$	1
				$CsBe_2F_5$	$P4_132$	2
				H_2O_2	$P4_12_12$	3
(10, 3)-*a* (90°)	$I2_13$			α-$Hg_3S_2Cl_2$	$I2_13$	4
				Sn_2F_3Cl*	$P2_13$	5
(10, 3)-*b* (120°)	$I4_1/amd$	8(*e*) $(00\frac{1}{12})$	$2\sqrt{3}$	α-$ThSi_2$	$I4_1/amd$	6
				P_2O_5	$Fdd2$	7
				$(Zn_2Cl_5)(H_5O_2)$	$Fdd2$	8
				$(CH_3)_2S(NH)_2$	$Fdd2$	9
				[GeS_2	$Fdd2$	10]
(10, 3)-*b* (90°)	$C2/c$			$La_2Be_2O_5$	$C2/c$	11
(10, 3)-*c*	$P3_112$	6(*c*) $(\frac{1}{3}\frac{1}{6}\frac{1}{9})$	$3\sqrt{3}/2$	B_2O_3	$P3_1$	12
(10, 3)-*d*	$Pnna$	8(*e*)†		α-resorcinol	$Pn2_1a$	13
				β-resorcinol		14
(8, 3)-*a*	$P6_222$	6(*i*) $(\frac{2}{5}\frac{4}{5}0)$	$3\sqrt{2}/5$	Caryophyllene chlorohydrin	$P3_1$	15
(8, 3)-*b*	$R\bar{3}m$	18(*f*)‡ $(\frac{2}{5}00)$	$\sqrt{6}/5$	$C_6H_4\{Si(CH_3)_2OH\}_2$	$P\bar{1}$	16
6.10^2	$R\bar{3}m$	18(*f*)†‡		$N_4(CH_2)_6.6H_2O$§	$R3m$	17
				β-quinol‖	$R\bar{3}$, $R3$	18

* In $(Sn_2F_3)Cl$ each Sn atom forms three bonds to bridging F atoms. Another study (private communication from J. D. Donaldson) shows that $(Sn_2F_3)Cl$ is actually orthorhombic, space group $P2_12_12_1$, pseudocubic, $a = b = c = 7.880$ Å.

† No unique configuration is possible with all links of equal length and all interbond angles equal to 120°.

‡ Hexagonal setting.

§ Of the three space groups $R32$, $R3m$, and $R\bar{3}m$, only $R3m$ is compatible with the point symmetry ($3m$) of the hexamethylene tetramine molecule, which is situated in the interstices of the 6.10^2 net formed by the hydrogen-bonded water molecules.

‖ This structure consists of two identical interpenetrating nets that enclose foreign molecules in the interstices. The symmetry appears to depend on the nature of the enclosed molecules. In $C_6H_4(OH)_2.\frac{1}{3}SO_2$ the SO_2 molecule becomes effectively centrosymmetric by rotation (space group $R\bar{3}$), but in $C_6H_4(OH)_2.\frac{1}{3}CH_3OH$ the methanol molecule apparently cannot rotate about a horizontal axis, and the space group is $R3$.

REFERENCES

1. K. H. Janson, H. Schäfer, and A. Weiss, *Z. anorg. Chem.*, 1970, **372**, 87; G. E. Pringle, *Acta Crystallogr.*, 1972, **B28**, 2326.
2. Y. Le Fur and S. Aléonard, *Acta Crystallogr.*, 1972, **B28**, 2115.
3. S. C. Abrahams, R. L. Collin, and W. N. Lipscomb, *Acta Crystallogr.*, 1951, **4**, 15.

Table 13.2 (*Continued*)

4. A. J. Frueh and N. Gray, *Acta Crystallogr.*, 1968, **B24**, 156.
5. G. Bergerhoff and L. Goost, *Acta Crystallogr.*, 1974, **B30**, 1362.
6. G. Brauer and A. Mitius, *Z. anorg. Chem.*, 1942, **249**, 325.
7. H. C. J. de Decker, *Rec. Trav. Chim. (Pays-Bas)*, 1941, **60**, 413.
8. H. Follner, *Acta Crystallogr.*, 1970, **B26**, 1544.
9. E. Prince, *Acta Crystallogr.*, 1975, **B31**, 2536.
10. W. H. Zachariasen, *J. Chem. Phys.*, 1936, **4**, 618.
11. L. A. Harris and H. L. Yakel, *Acta Crystallogr.*, 1968, **B24**, 672.
12. S. L. Strong and R. Kaplow, *Acta Crystallogr.*, 1968, **B24**, 1032.
13. J. M. Robertson, *Proc. Roy. Soc. (A)*, 1936, **157**, 79.
14. J. M. Robertson and A. R. Ubbelohde, *Proc. Roy. Soc. (A)*, 1938, **167**, 122.
15. D. Rogers and Mazhar-ul-Haque, *Proc. Chem. Soc.*, 1963, 371.
16. L. E. Alexander, M. G. Northolt, and R. Engmann, *J. Phys. Chem.*, 1967, **71**, 4298.
17. T. C. W. Mak, *J. Chem. Phys.*, 1965, **43**, 2799.
18. D. E. Palin and H. M. Powell, *J. Chem. Soc.*, 1947, 208.

Table 13.3 The net (10, 3)-*a*

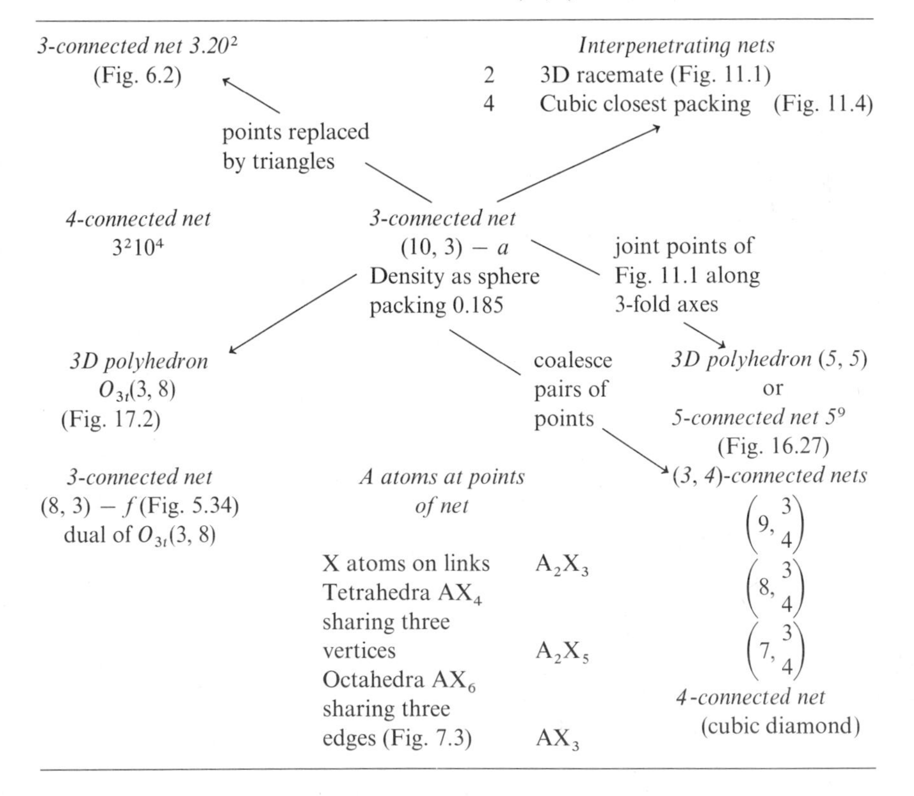

Three-dimensional Polyhedra

The preceding pages have described the basic topology and geometry of many nets, for the most part 3-, (3, 4)-, or 4-connected. We have also commented on some related topics such as the relation of nets to open sphere packings, to the closest packings of equal spheres, and to the formation of systems of interpenetrating nets. Next we describe tessellations of polygons on surfaces resulting from inflating the links of the basic nets—3D polyhedra. Tables 13.3 and 13.4 bring together these various aspects of our subject for the two simplest uniform 3-connected nets.

Table 13.4 The net (10, 3)-*b*

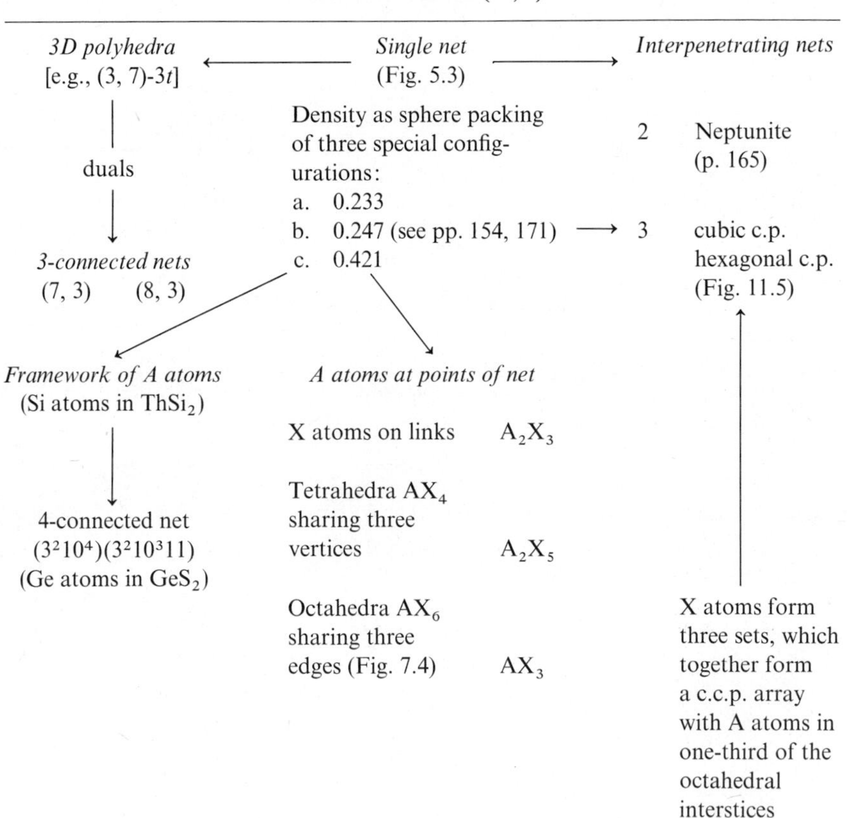

14

Appendix

Plane nets

Reference is made in a number of places to plane nets, which figure prominently in the derivation of some of the 3D nets. We therefore include a note on the simpler 3-, (3, 4)-, and 4-connected plane nets.

Plane nets conform to the following equations, in which ϕ_n is the fraction of the total number of polygons that are n-gons:

$$\text{3-connected:} \quad 3\phi_3 + 4\phi_4 + 5\phi_5 + 6\phi_6 + 7\phi_7 + \cdots + n\phi_n = 6 \qquad (14.1)$$

$$\text{4-connected:} \quad 3\phi_3 + 4\phi_4 + 5\phi_5 + 6\phi_6 + 7\phi_7 + \cdots + n\phi_n = 4 \qquad (14.2)$$

$$\text{5-connected:} \quad 3\phi_3 + 4\phi_4 + 5\phi_5 + 6\phi_6 + 7\phi_7 + \cdots + n\phi_n = \frac{10}{3} \qquad (14.3)$$

$$\text{6-connected:} \quad 3\phi_3 + 4\phi_4 + 5\phi_5 + 6\phi_6 + 7\phi_7 + \cdots + n\phi_n = 3 \qquad (14.4)$$

The special solutions, in which all the polygons are of one kind (n-gons) and all points are p-connected, are

$$\phi_6 = 1 \qquad \phi_4 = 1 \qquad \phi_3 = 1$$

In their most symmetrical configurations (with regular polygons) these are the three regular plane nets. There is clearly no solution of this kind of (14.3), and the solution $\phi_3 = 1$ is the *only* solution of (14.4).

The three regular plane nets are the 2D analogues of the regular solids. Corresponding to the Archimedean solids are the nets in which all points are

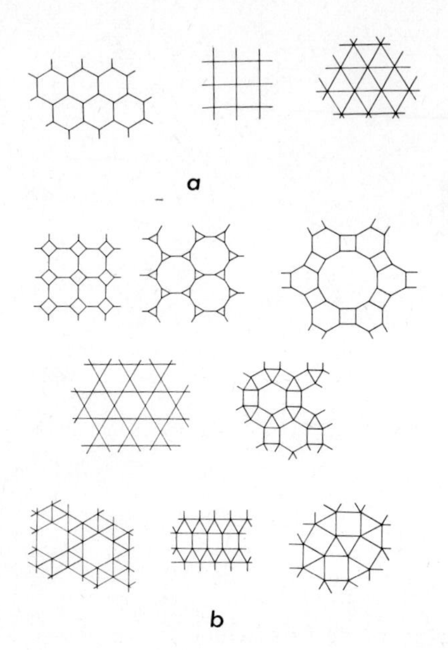

Fig. 14.1. Division of the plane into regular polygons (*a*) of the same kind and (*b*) of more than one kind.

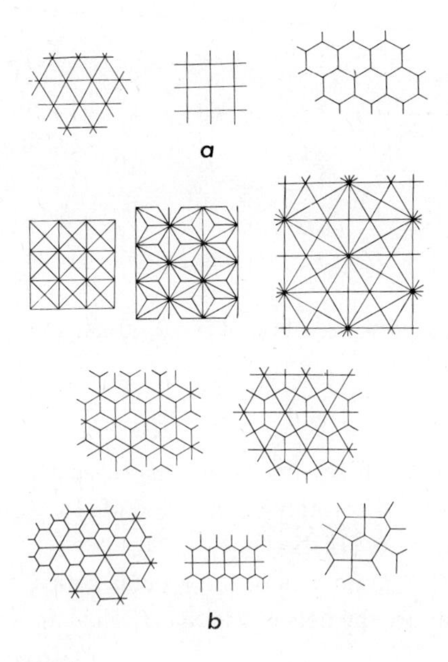

Fig. 14.2. Division of the plane into congruent polygons.

p-connected and topologically equivalent (i.e., have the same cyclic arrangement of polygons around each point) but contain polygons of two or more kinds; they include an additional eight nets (Fig. 14.1). The duals of these 8 nets correspond to the Catalan solids and represent the ways in which the plane can be divided into congruent polygons (Fig. 14.2). [There is an obvious (sheared) variant of the net $3^2 6^2$ with continuous rows of edge-sharing triangles and hexagons in which all the points are *not* equivalent; cf. the variants of the rhombicuboctahedron and icosidodecahedron.]

3-connected plane nets

Some of the simplest plane 3-connected nets are illustrated in Fig. 14.3 and listed in Table 14.1 in order of increasing "order" *m*, where *m* is the highest

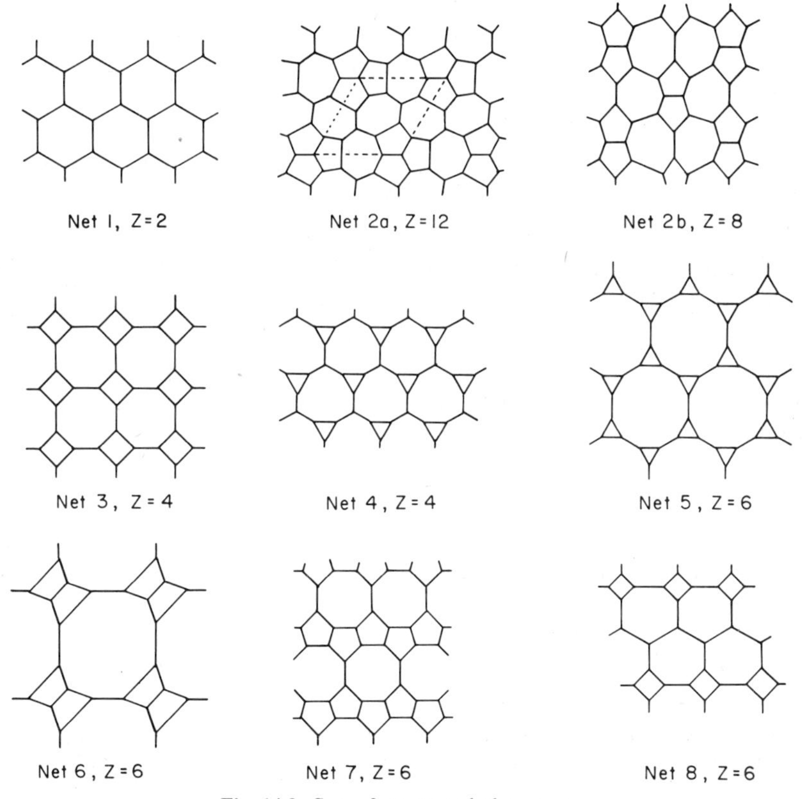

Fig. 14.3. Some 3-connected plane nets.

Table 14.1 3-Connected plane nets

m	Net	Z	ϕ_3	ϕ_4	ϕ_5	ϕ_6	ϕ_7	ϕ_8	ϕ_9	ϕ_{10}	ϕ_{11}	ϕ_{12}
1	1	2	—	—	—	1	—	—	—	—	—	—
2	2	8, 12	—	—	$\frac{1}{2}$	—	$\frac{1}{2}$	—	—	—	—	—
—	3	4	—	$\frac{1}{2}$	—	—	—	$\frac{1}{2}$	—	—	—	—
—	4	4	$\frac{1}{2}$	—	—	—	—	—	$\frac{1}{2}$	—	—	—
3	5	6	$\frac{2}{3}$	—	—	—	—	—	—	—	—	$\frac{1}{3}$
—	6	6	—	$\frac{2}{3}$	—	—	—	—	—	$\frac{1}{3}$	—	—
—	7	6	—	—	$\frac{2}{3}$	—	—	$\frac{1}{3}$	—	—	—	—
—	8	6	—	$\frac{1}{3}$	—	—	$\frac{2}{3}$	—	—	—	—	—
—	9	6	$\frac{1}{3}$	$\frac{1}{3}$	—	—	—	—	—	—	$\frac{1}{3}$	—
—	10	6	$\frac{1}{3}$	—	$\frac{1}{3}$	—	—	—	—	$\frac{1}{3}$	—	—
—	11	12	$\frac{1}{3}$	—	—	$\frac{1}{3}$	—	—	$\frac{1}{3}$	—	—	—
—	12	6	$\frac{1}{3}$	—	—	—	$\frac{1}{3}$	$\frac{1}{3}$	—	—	—	—
—	13	12	—	$\frac{1}{3}$	$\frac{1}{3}$	—	—	—	$\frac{1}{3}$	—	—	—
—	14	6	—	$\frac{1}{3}$	—	$\frac{1}{3}$	—	$\frac{1}{3}$	—	—	—	—
—	15	12	—	—	$\frac{1}{3}$	$\frac{1}{3}$	$\frac{1}{3}$	—	—	—	—	—

denominator in ϕ_n. For 3-connected nets the number Z of points in the repeat unit is $2m$ or a multiple of $2m$. For example, there is no configuration of Net 2 of Table 14.1 with 4 points in the repeat unit but there are configurations with 8 and 12 points in their repeat units. Two features of these nets should be noted. (*a*) Specifying the values of ϕ_n does not uniquely define a plane net in which there are polygons of more than one kind, for there is an indefinitely large number of different relative arrangements of the polygons in the same relative proportions. (*b*) Moreover, two or more such nets differing in the arrangement but not the proportions of the polygons of different kinds may have the same value of Z. These points are discussed further in Part 9 for the next simplest set of solutions of (14.1), namely,

$$\phi_5 = \phi_7 = \tfrac{1}{2} \qquad \phi_4 = \phi_8 = \tfrac{1}{2} \qquad \phi_3 = \phi_9 = \tfrac{1}{2}$$

and both are illustrated for the second of these solutions in Fig. 14.4. Note also that only one of the configurations of Fig. 14.4, namely the case of $Z = 4$, has the *point symbol* 4.8^2, which indicates that one 4-gon and two 8-gons meet at every point.

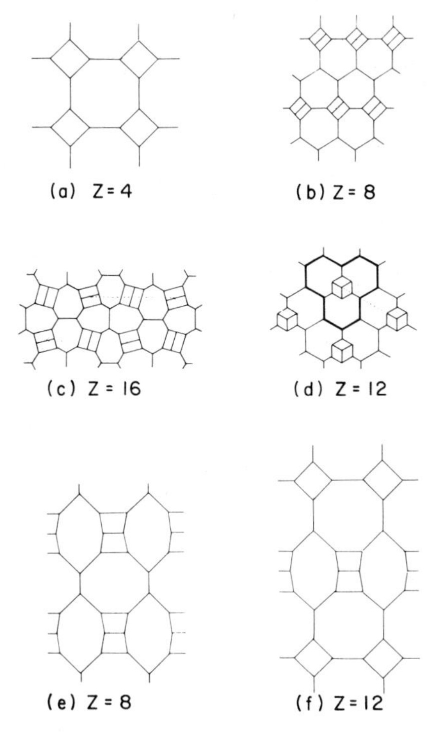

Fig. 14.4. Configurations of the 3-connected net $\phi_4 = \phi_8 = \frac{1}{2}$.

4-connected plane nets

The first six solutions of (14.2) are listed in Table 14.2, and the nets are illustrated, some in more than one configuration, in Fig. 14.5.

Table 14.2 4-Connected plane nets

m	Net	Z	ϕ_3	ϕ_4	ϕ_5	ϕ_6	ϕ_7
1	1	1	—	1	—	—	—
2	2	4	$\frac{1}{2}$	—	$\frac{1}{2}$	—	—
3	3	6, 9	$\frac{1}{3}$	$\frac{1}{3}$	$\frac{1}{3}$	—	—
	4	3	$\frac{2}{3}$	—	—	$\frac{1}{3}$	—
4	5	12	$\frac{3}{4}$	—	—	—	$\frac{1}{4}$
	6	4	$\frac{1}{2}$	$\frac{1}{4}$	—	$\frac{1}{4}$	—

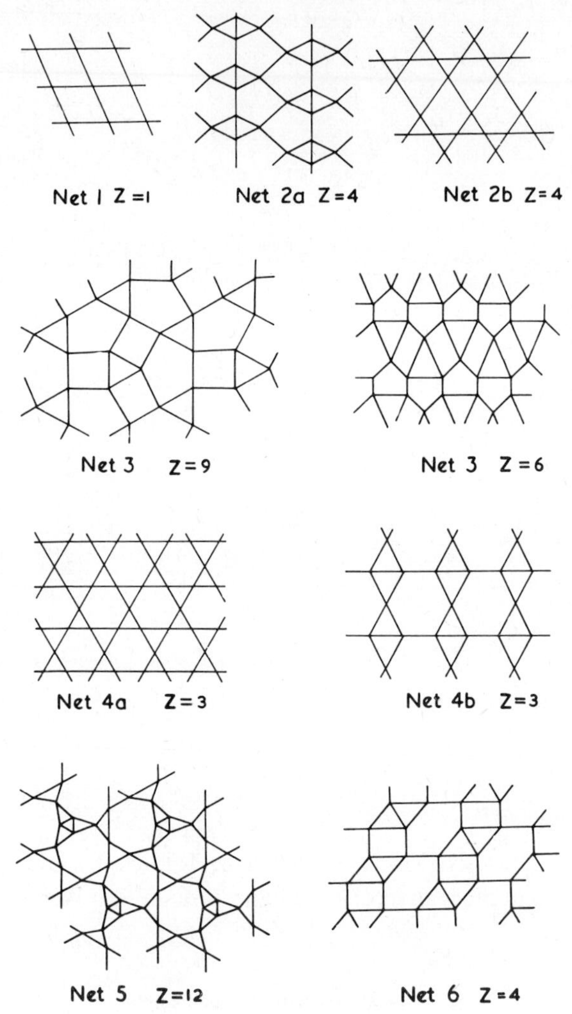

Fig. 14.5. Some 4-connected plane nets.

(3, 4)-Connected plane nets

For these nets there is no single equation similar to (14.1)–(14.4) because the value of $\sum n\phi_n$ depends on the ratio of 3- to 4-connected points. It can be shown that

$$\sum n\phi_n = \frac{2(3r + 4)}{r + 2}$$

where r is the ratio of the numbers of 3- to 4-connected points. The simplest form of this equation arises when $r = 2$:

$$\sum n\phi_n = 5$$

with the special solution $\phi_5 = 1$. The two forms of this plane net (which cannot be realized with regular pentagons) are illustrated in Fig. 14.6.

In a periodic (3, 4)-connected net, as also in a convex polyhedron with only 3- and 4-connected vertices, the number of 3-connected vertices in a repeat unit must be even. The minimum value of Z is therefore 3. In the derivation of 4-connected 3D nets we refer to (3, 4)-connected plane nets having $Z = 3$

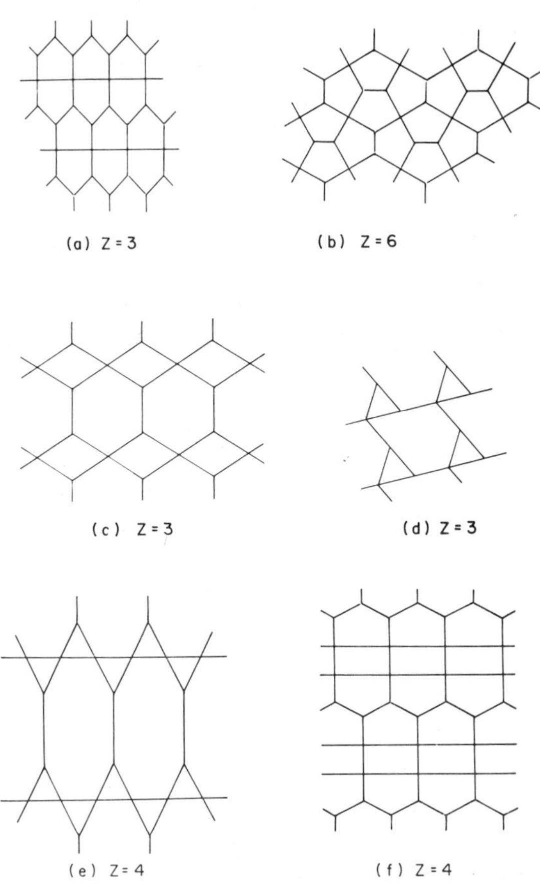

Fig. 14.6. Configurations of (3, 4)-connected nets. (*a*)–(*d*) $\sum n\phi_n = 5$; (*e*), (*f*) $\sum n\phi_n = 14/3$ (see also Fig. 9.2).

and $Z = 4$. These correspond to $r = 2$ and 1. For $r = 2$, $\sum n\phi_n = 5$, as already noted, and the simplest solutions are as follows

$$\phi_5 = 1 \qquad \text{Fig. 14.6}a \text{ and } b$$

$$\phi_4 = \phi_6 = \tfrac{1}{2} \qquad \text{Fig. 14.6}c$$

$$\phi_3 = \phi_7 = \tfrac{1}{2} \qquad \text{Fig. 14.6}d$$

Each solution has a configuration with $Z = 3$. For $r = 1$, $\sum n\phi_n = \frac{14}{3}$. Nets with $Z = 4$ used to derive 4-connected 3D nets were listed in Table 9.1 and illustrated in Fig. 9.2, but it was noted that other configurations of two of the nets have subsequently been found; they are included in Fig. 14.6.

PART II

Three-dimensional polyhedra

15

Finite and infinite polyhedra

First we remind the reader of the equations relating to finite (simply connected) convex polyhedra. From Euler's relation between the numbers of vertices (N_0), edges (N_1), and faces (N_2)

$$N_0 - N_1 + N_2 = 2$$

the following equations can be derived for the special families of polyhedra that have three, four, or five edges meeting at each vertex and f_n n-gon faces:

$$\text{3-connected:} \quad 3f_3 + 2f_4 + f_5 \pm 0f_6 - f_7 - 2f_8 - \cdots = 12 \quad (15.1)$$

$$\text{4-connected:} \quad 2f_3 \pm 0f_4 - 2f_5 - 4f_6 - \cdots = 16 \quad (15.2)$$

$$\text{5-connected:} \quad f_3 - 2f_4 - 5f_5 - 8f_6 - \cdots = 20 \quad (15.3)$$

These equations are of the general form

$$\sum f_n\{4 - (n-2)(p-2)\} = 4p \quad (15.4)$$

and their special solutions (all faces of the same type) correspond to the five Platonic solids, namely,

3-connected: $f_3 = 4$, $f_4 = 6$, $f_5 = 12$

4-connected: $f_3 = 8$

5-connected: $f_3 = 20$

In the analogous equation for 6-connected polyhedra, the coefficient of f_3 is zero and all the other coefficients have negative values. There is therefore no simple convex polyhedron having six (or more) edges meeting at *every* vertex. For $p > 6$ the coefficients of all f_n are negative and the equations analogous to (15.1)–(15.3) do not have solutions realizable as finite convex polyhedra.

The foregoing proof of the existence of only five convex polyhedra (n, p) is purely topological. It happens that these are also the only convex polyhedra (n, p) that can be constructed with *regular plane* faces. This correspondence between the number of topological entities (n, p) and the number of geometrical entities (n, p) does not hold for the 3D polyhedra we shall describe. It is therefore convenient to use the term "Platonic" or "uniform" to describe a system (n, p) and to restrict the term "regular" to polyhedra that can be constructed with regular plane faces.

An infinite polyhedron is formed by inflating the links of a 2D or 3D net to form tunnels that meet at the nodes of the net, then inscribing a tessellation of polygons on the surface so formed. All examples of 3D polyhedra to be described are based on one or other of the nets illustrated in Fig. 15.1 plus the dual of Fig. 15.1*d*, the *I* lattice; in these nets all points are equivalent and p-connected. Instead of the single solution for each of the systems (3, 3), (3, 4), (3, 5), and (3, 6), we find that the tessellation (3, 8), for example, can be inscribed on the surfaces derived from all the nets of Fig. 15.1 (and also the *I* lattice). Moreover there may be more than one surface of the same topological type (e.g., 6-tunnel) on which a given tessellation (n, p) can be inscribed; see (3, 8)-6*t* and the summarizing Table 19.1.

We deal now with the more general case of a 3D surface and note later two examples of infinite polyhedra based on 2D nets. The curvature varies over such a surface and the links (sides of polygons) are curved lines connecting the points on the surface. Unless otherwise stated, we restrict ourselves to tessellations (n, p), that is, to polyhedra having all faces n-gons, of which p meet at each point (vertex). These can be described as uniform 3D polyhedra, or homologues of the Platonic solids. In constructing models we have made all links equal in length. All polygons are therefore equilateral, but owing to the varying curvature, they are not all equiangular. We also consider later which of the 3D polyhedra can be constructed with *plane* faces, that is, with regular plane n-gons; these are the homologues of the *regular* solids.

When n is small there is a clearly recognizable surface enclosing the 3D polyhedron. The space not enclosed by the surface, the space between the

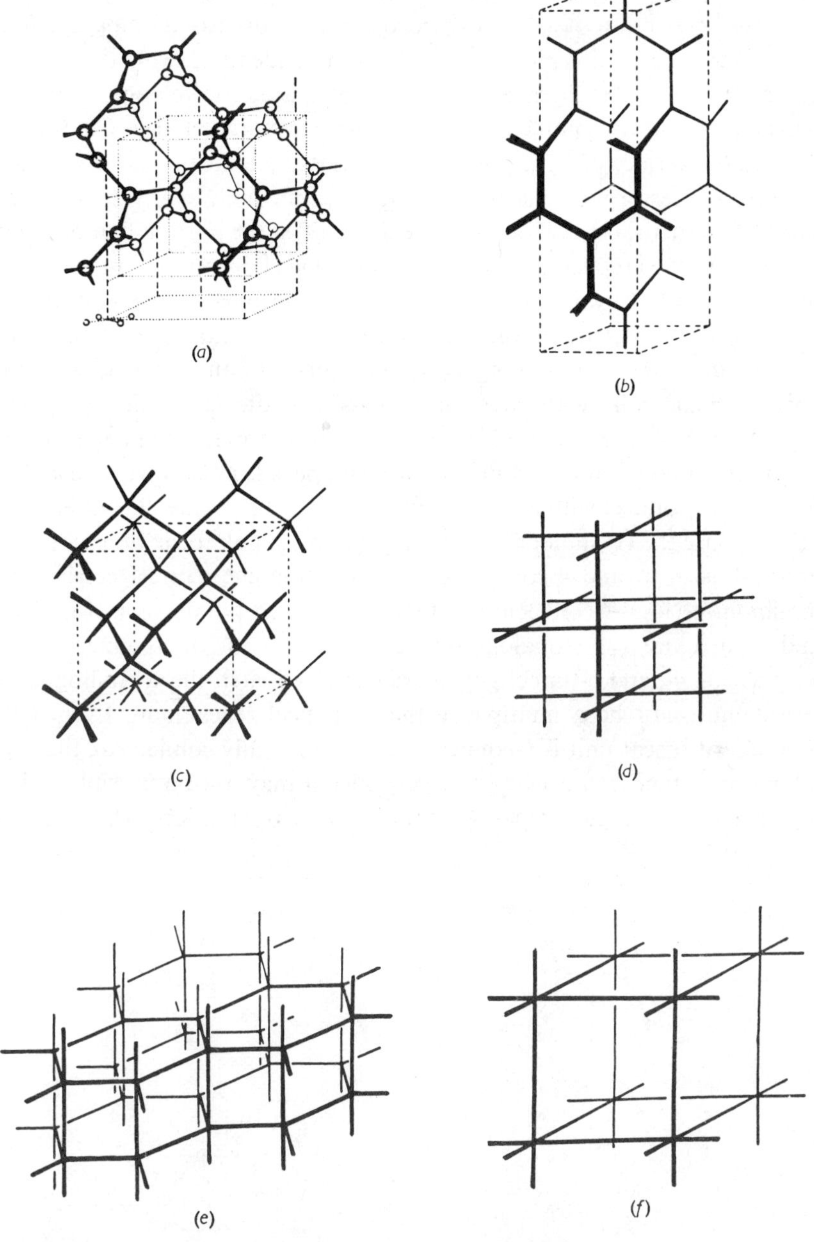

Fig. 15.1. Some 3D nets. (*a*), (*b*) 3-Connected; (*c*), (*d*) 4-connected; (*e*) 5-connected; (*f*) 6-connected.

tunnels, is described as the *complementary polyhedron* if it conforms to the previous description of a 3D polyhedron, a point discussed on page 250. In some cases the complementary polyhedron is identical with the original polyhedron, when each encloses one-half of space. The relation between a polyhedron and its complement is examined in more detail in Chapter 18.

Since the surface of a 3D polyhedron repeats periodically in three dimensions, it is necessary to consider only the repeating unit. In any three-dimensional pattern, the crystallographic "repeat unit" is the portion that produces the pattern when repeated *in the same orientation* at distances consistent with one of the 14 Bravais lattices. This crystallographic repeat unit can be distinguished from the topological repeat unit. For example, in the plane (6, 3) net all points are topologically equivalent, and the topological repeat unit is a single point with three "half-bonds" as indicated by the heavy lines Fig. 15.2*a*. The crystallographic repeat unit is, however, the system of two points enclosed within the unit cell (broken lines). It must have at least four free links to connect with four other units to form an infinite 2D pattern. In a 3D structure the crystallographic repeat unit must link to six other identical units; thus in 3- and 4-connected nets, for example, it must consist of a minimum of four 3-connected or two 4-connected points, as in Fig. 15.2*b* and *c*. Therefore if the topological repeat unit is a 3-, 4-, or 5-connected unit, as in the 3-, 4-, and 5-tunnel polyhedra described later, the crystallographic repeat unit *must* be a multiple of the toplogical repeat unit. Even if the topological repeat unit is 6-connected (or more highly connected), the crystallographic repeat unit *may* be larger, since it may not be possible to join together the topological repeat units in the same orientation. (This is true of

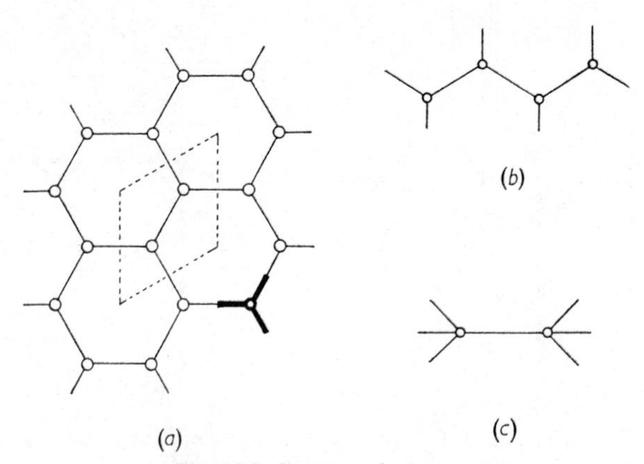

Fig. 15.2. Repeat units in nets.

the 6-tunnel (3, 8) unit of Fig. 16.8*b*.) In what follows the term "repeat unit" normally refers to the topological repeat unit.

Part 7 included the following derivation of a modified form of the Euler relation that is applicable to 3D polyhedra and gives an expression for Z, the number of points in the repeat unit of a 3D polyhedron, in terms of n, p, and t (the number of tunnels).

A repeat unit may be dissected out of an infinite 3D polyhedron by cutting around each tunnel *along links of the tessellation*, to give a polyhedral unit with t faces (holes) representing the tunnels, as shown in Fig. 15.3 for a 6-tunnel unit. To this finite "polyhedron" we may apply Euler's relation, $N_0 - N_1 + N_2 = 2$. The values of N_0, N_1, and N_2 all differ from the corresponding quantities for the repeat unit of the polyhedral surface, whose values can be written Z, E, and F. Clearly $N_2 = F + t$. Since the t faces are shared when the units are joined together, the corners and edges of their faces count as only half-points or half-edges for the repeat unit but as whole points or edges for the finite polyhedron. Also, since a polygon has the same number of edges as corners, the number of shared points is the same as the number of shared edges. The excess of edges over vertices therefore remains the same for both the connected and unconnected units, so that $Z - E = N_0 - N_1$. Hence

$$Z - E + F + t = 2 \qquad \text{or} \qquad Z - E + F = 2 - t$$

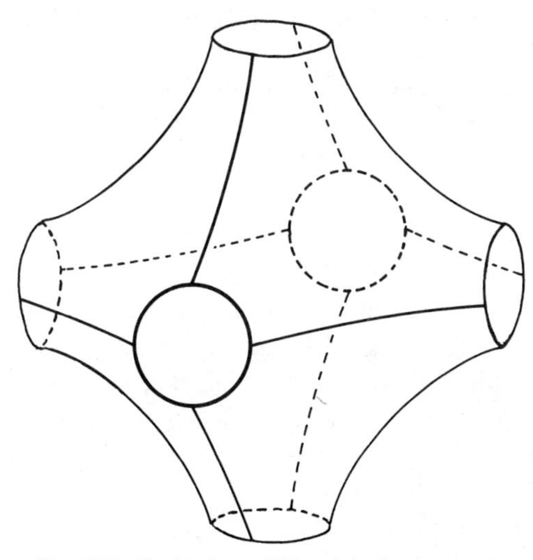

Fig. 15.3. Derivation of 3D polyhedra (see text).

Since p n-gons meet at every point of the tessellation and every n-gon has n vertices, the ratio of n-gons to points is p/n, or $F = Zp/n$. Since p edges meet at each point and each edge connects two points, the ratio of edges to points is $p/2$, or $E = Zp/2$. Substitution in the modified Euler equation gives

$$Z = \frac{2n(2-t)}{2p+2n-np} \quad \text{or} \quad \frac{2n(2-t)}{4-(n-2)(p-2)}$$

$$E = \frac{np(2-t)}{4-(n-2)(p-2)} \qquad F = \frac{2p(2-t)}{4-(n-2)(p-2)}$$

Comparison of these expressions with those for a finite polyhedron, namely,

$$N_0 = \frac{4n}{4-(n-2)(p-2)} \qquad N_1 = \frac{2np}{4-(n-2)(p-2)}$$

$$N_2 = \frac{4p}{4-(n-2)(p-2)}$$

or alternatively consideration of the derivation of (15.1)–(15.3) shows that the generalized form of (15.4) is

$$\sum f_n\{4-(n-2)(p-2)\} = 2p(2-t) \tag{15.5}$$

The values of Z derived from the expression just given agree with those observed for all the 3D polyhedra to be described. However examination of these polyhedra revealed that this derivation does not appear to be strictly applicable to many of the types of polyhedra involved. Figure 15.3 implies that the 3D polyhedron is made of finite "polyhedral" repeat units that share faces, of which there is presumably an even number. The number of tunnels (the connectedness p of the underlying net) is not always even, however. In some 3D polyhedra, moreover, the tunnels themselves are polyhedral (e.g., tetrahedral or octahedral in triangulated nets), so that additional edges and/ or faces are in fact inserted between the units of the simple type illustrated in Fig. 15.3. It would seem that three cases should be recognized (Fig. 15.4), of which (i) corresponds to Fig. 15.3, one face being shared between each pair of adjacent polyhedral repeat units (R_1 and R_2). In (ii) R_1 and R_2 share an edge, and comparing the numbers of vertices, edges, and faces of R_1 (or R_2) with the corresponding numbers for the repeat unit of the resulting 3D polyhedron we find

$$Z = N_0 - t \qquad E = N_1 \qquad F = N_2$$

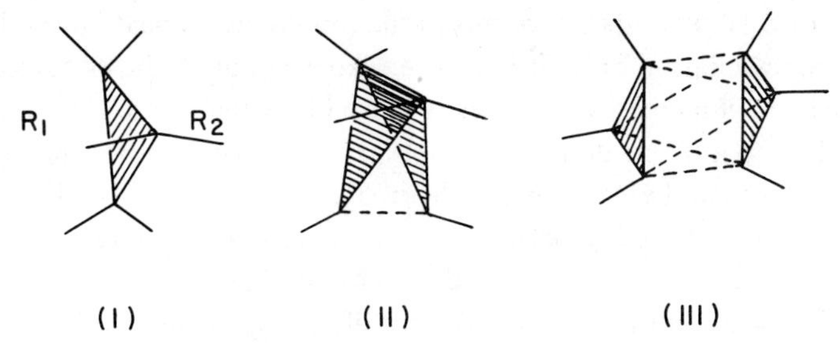

Fig. 15.4. Relation between adjacent polyhedral repeat units (R_1 and R_2) in 3D polyhedra.

Substituting in $N_0 - N_1 + N_2 = 2$, we have $Z - E + F = 2 - t$ as for case i. In iii the formation of "octahedral" tunnels leads to the following relations between R_1 and the repeat unit of the 3D polyhedron:

$$Z = N_0 \qquad E = N_1 + 3t \qquad F = N_2 + 2t$$

whence $Z - E + F = 2 - t$, as before. All three cases therefore lead to the same expression for Z, in agreement with observation.

We have supposed that a tunnel connects only two units (2-valent tunnel), that is, it is topologically equivalent to a *link* of a 3D net. We could generalize the treatment by introducing multivalent tunnels connecting three or more units. An example would be a tetrahedral tunnel sharing its faces with four units, giving

$$Z = \frac{n(4 - 3t)}{4 - (n - 2)(p - 2)}$$

For $n = p = 6$ and $t = 4$, this gives $Z = 4$, corresponding to Coxeter's $\{6, 6/3\}$, in which truncated tetrahedra and tetrahedra are situated at alternate nodes of the diamond net; each of the 12 vertices of the *tt* is common to three *tt*. However, this is the only known example of such a polyhedron, and it could be described equally well as being built from repeat units of two kinds.

Note on Nomenclature

Since we are using the same type of symbol (n, p) for both 3D polyhedra and uniform 3D nets, it is important to note the difference between the meanings of the symbol in the two cases. The symbol (n, p) for a 3D polyhedron (hence the value of Z_t, the number of points in the topological repeat unit) relates to the surface tessellation only; it does not take account of the circuits *around* the tunnels. As n increases, it becomes more likely that there will be circuits smaller than n-gons around the tunnels. This is true, for example, for one (6, 4) polyhedron and for certain polyhedra (7, 3), (8, 3), (9, 3), and (12, 3) which are the duals of polyhedra (3, p) built with "octahedral" tunnels, for around the tunnels of these polyhedra there are circuits of 6 links. If we follow the apparently logical path and take account of *all* circuits in any 3D system of Table 1.1, these polyhedra are not to be described as nets (n, p) because they contain circuits smaller than n-gons.

Solutions of Equation (15.5)

We now consider the solutions of this equation for various values of t, namely,

$t = 0$	finite polyhedra, that is, equations (15.1)–(15.3), which we need not discuss further
$t = 1$	tessellations on a cylindrical surface that is closed at one end; these are more conveniently considered after $t = 2$
$t = 2$	periodic plane nets
$t \geqslant 3$	3D polyhedra

Periodic plane nets: $t = 2$

Putting $p = 3$ and $t = 2$ in (15.5) we have instead of (15.1):

$$3f_3 + 2f_4 + f_5 \pm 0f_6 - f_7 - \cdots = 0 \tag{15.6}$$

Since the right-hand side is zero, we may replace f_n by ϕ_n, where ϕ_n is the *fraction* of polygons that are n-gons. Subtracting $6\sum\phi_n = 6$ and changing all signs, we have

$$3\phi_3 + 4\phi_4 + 5\phi_5 + 6\phi_6 + 7\phi_7 + \cdots = 6 \tag{15.7}$$

By similar manipulations of the equations for $p = 4$, 5, and 6, we find

$$\sum n\phi_n = 4 \qquad \text{for} \quad p = 4$$

$$\sum n\phi_n = \tfrac{10}{3} \qquad \text{for} \quad p = 5$$

$$\sum n\phi_n = 3 \qquad \text{for} \quad p = 6$$

These are the familiar equations for plane nets, of which the special solutions are

$$\left.\begin{array}{ll} p = 3 & \phi_6 = 1 \\ p = 4 & \phi_4 = 1 \\ p = 6 & \phi_3 = 1 \end{array}\right\} \text{Regular plane nets}$$

The plane nets that can be drawn with regular polygons of two or more kinds (semiregular plane nets) are the analogues of the Archimedean solids, and their duals are the analogues of the Catalan solids; they were illustrated in Figs. 14.1 and 14.2.

Tessellations on a cylinder closed at one end; $t = 1$

Strips (of various widths) of plane nets can be wrapped around the surface of a cylinder that is open at both ends, provided the cuts are made in appropriate positions. Such tessellations can be regarded as 2-tunnel systems. The solutions of (15.5) for $t = 1$ correspond to the combinations of polygons required to fill in one end of a cylinder on which is inscribed one of the regular nets. It suffices to consider the form of (15.5) for $p = 3$, $t = 1$:

$$3f_3 + 2f_4 + f_5 \pm 0f_6 - f_7 - \cdots = 6 \qquad (15.8)$$

The tessellation to be inscribed on the cylindrical surface is then the regular 3-connected net (6, 3).

We first note that this net can be derived by starting from a central point and surrounding it by successive shells of 3, 9, 15, and so on, hexagons (Fig. 15.5*a*) or from a central hexagon surrounded by successive shells of 6, 12, 18, and so on, hexagons (Fig. 15.5*b*). This arithmetic increase in the number of hexagons in successive shells leads to a *periodic* plane net. An alternative way of illustrating the nets of Fig. 15.5*a*, *b* is shown in Fig. 15.5*c*, *d*; all represent the same infinite periodic plane net (6, 3).

If we wish to produce a diagram of this type representing a 6-gon net wrapped around a cylinder that is closed at one end (the center of the diagram), we must have the same number of hexagons in successive shells, this number (w) being the number of hexagons in the strip of (6, 3) net. Thus if the strip of net of Fig. 15.6 is wrapped around a cylinder so that edges drawn as heavy lines coincide, we need three hexagons in each successive shell instead of 3, 9, 15, and so on, or 6, 12, 18, and so on. The simplest solutions of (15.8) are listed in Table 15.1 together with the width (w) of 6-gon net, and some are illustrated in Fig. 15.7.

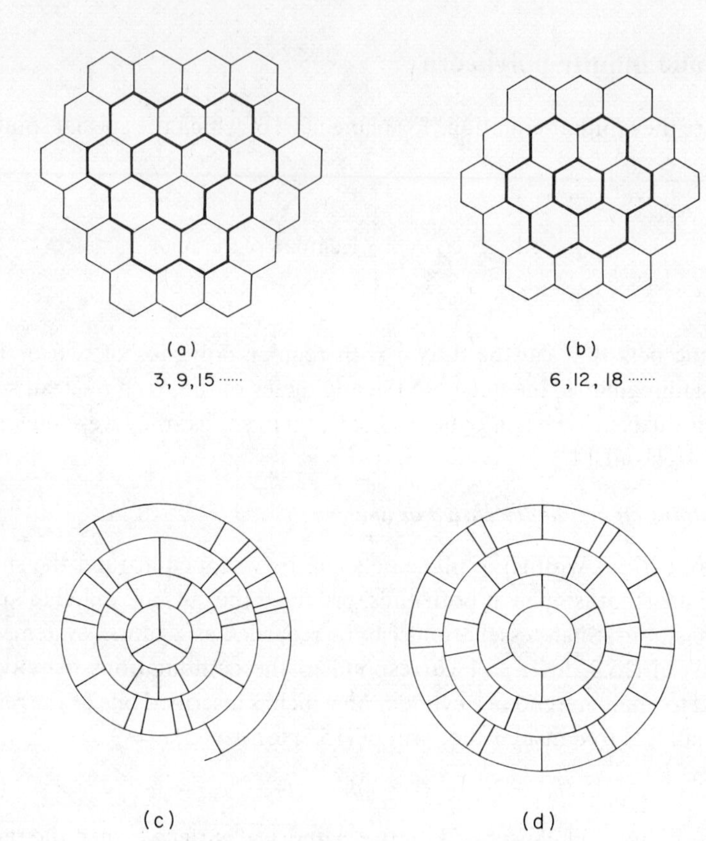

Fig. 15.5. The plane (6, 3) net shown as (*a*) successive shells of 3, 9, 15, . . . , 6-gons around a central point and (*b*) shells of 6, 12, 18, . . . , 6-gons around a central 6-gon. The alternative representations (*c*) and (*d*) are equivalent to (*a*) and (*b*), respectively.

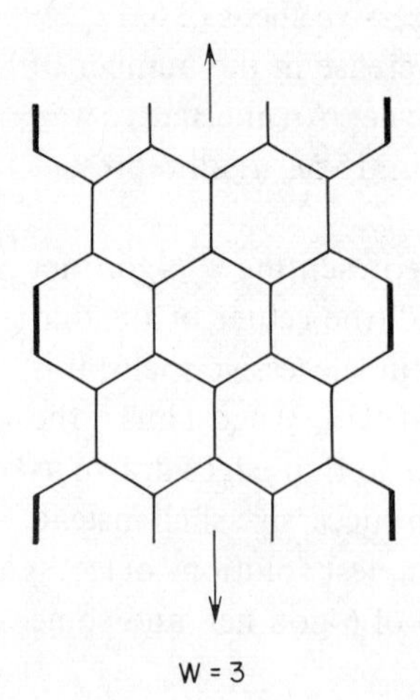

Fig. 15.6. Strip of the plane (6, 3) net, three 6-gons wide, that can be wrapped around a cylinder so that the links drawn as heavy lines coincide.

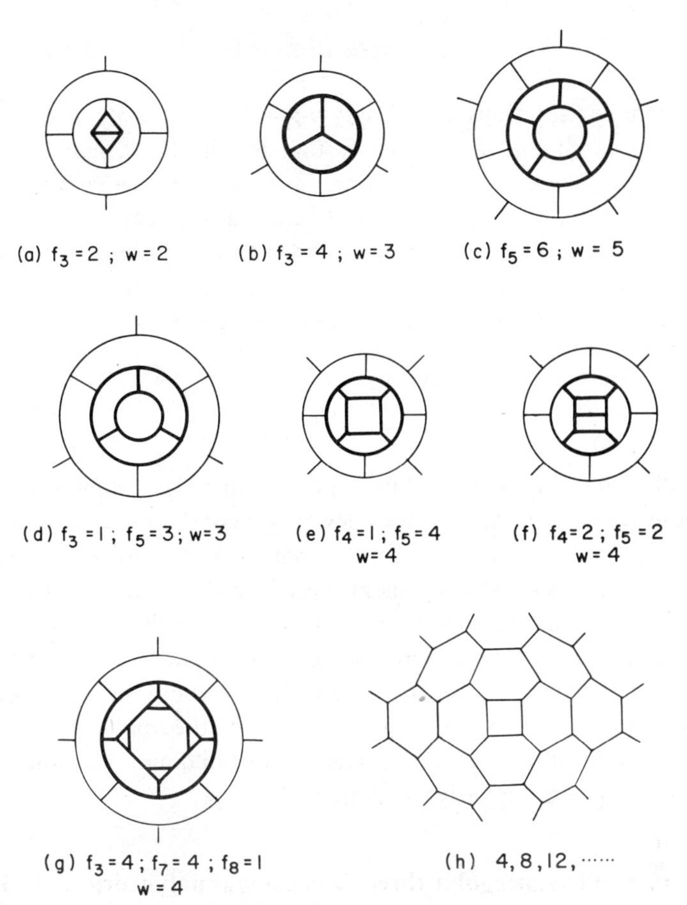

Fig. 15.7. Topological diagrams of tessellations on a cylinder closed at one end. Heavier lines enclose the solutions of (15, 8), beyond which all polygons are 6-gons.

Table 15.1 Solutions of equation (15.8)

f_3	f_4	f_5	w
2	—	—	2
—	3	—	3, 4
—	—	6	5
1	—	3	3
1	1	1	3, 4
—	1	4	4
—	2	2	4

More complex solutions include any n-gon $(n > 6) + nf_5$, and an indefinite number of combinations such as that in Fig. 15.7*g*, namely, $f_3 = 4$, $f_7 = 4$, $f_8 = 1$. These solutions are obviously equal to one-half the values of f_n for certain 3-connected polyhedra. By closing both ends of the cylinder, we form a polyhedron that has been extended by interposing the 6-gon faces we have drawn on the cylinder. The addition of any number of 6-gon faces is permitted by (15.1), since the coefficient of f_6 is zero, as in

$$f_3 = 4 \text{ (tetrahedron)} \qquad \text{and} \qquad \left.\begin{matrix} f_3 = 4 \\ f_6 = 4 \end{matrix}\right\} \text{Truncated tetrahedron}$$

though all such solutions are not necessarily realizable as convex polyhedra.

We note in passing that if we insert fewer polygons than those corresponding to a solution of (15.8), we can form tessellations that also have arithmetically increasing numbers of 6-gons in successive shells; an example is given in Fig. 15.7*h*. If constructed with equal links, this forms a tessellation on a bowl-shaped surface, intermediate between the cylindrical and plane nets. When a large portion of a net such as this is spread out on a flat surface so that the disturbance to the 6-gon net is confined to the small area around the 4-gon, the model illustrates how a mistake in building a structure can be accommodated within a fairly small area.

Regular and semiregular three-dimensional polyhedra: $t \geqslant 3$

The results of generalizing (15.4) to (15.5) are as follows.

1. For $p \geqslant 6$ the equations have positive solutions for $t \geqslant 3$ correspond ing to the numbers and types of polygons forming the surfaces of infinite 3D polyhedra, f_n now being the number of n-gon faces in the repeat unit. For example, the expanded form of (15.5) for $p = 7$ is

$$-f_3 - 6f_4 - 11f_5 - 16f_6 - \cdots = 14(2 - t) \qquad (15.9)$$

which for $t = 3$ becomes

$$f_3 + 6f_4 + 11f_5 + 16f_6 + \cdots = 14 \qquad (15.10)$$

an equation having the special solution $f_3 = 14$ for a surface tessellation consisting of triangles only.

2. Whereas each of the equations (15.1)–(15.3) has only one solution for f_3, for example, each of the equations such as (15.9) has a set of solutions,

one for each value of t, all having $n = 3$ and $p = 7$:

t:	3	4	5	6 etc.
f_3:	14	28	42	56

Moreover, there may be more than one way of constructing a polyhedron with specified values of n, p, and t; that is, there may be more than one t-tunnel unit for a given tessellation (n, p): see Table 19.1. We could therefore draw up a table for each value of n; Table 15.2 is for the triangulated polyhedra ($n = 3$). There is no reason to suppose that all these polyhedra could be constructed, for it is not possible to construct all finite convex polyhedra that are consistent with the simple Euler relation (e.g., the 3-connected polyhedron $f_5 = 12$, $f_6 = 1$). We examine systematically the families (n, p) in the next section.

Table 15.2 Number of faces in topological repeat units of (3, *p*) polyhedra*

	t									
p	3	4	5	6	7	8	9	10	11	12
7	14	**28**	42	56	70	84	98	112	126	140
8	**8**	**16**	24	**32**	40	48	56	64	72	80
9	6	**12**	18	24	30	**36**	42	48	54	60
10	†	10	†	**20**	†	30	†	40	†	50
11	‡	‡	‡	‡	22	‡	‡	‡	‡	44
12	4	8	12	16	20	**24**	28	32	36	40

* Boldface type distinguishes polyhedra realizable with plane equilateral triangular faces (see Table 17.1).
† Z nonintegral.
‡ Number of faces nonintegral.

The polyhedra that have all faces of the same kind (all n-gons) are the 3D homologues of the five Platonic solids, all of which can be constructed with plane regular faces. As already remarked, the 3D polyhedra (n, p) with curved faces and edges are much more numerous than the finite convex polyhedra, but only a limited number of these can be constructed with plane regular faces, as described in Chapter 17. Only the latter are the strict homologues of the *regular* solids, and of these the triangulated bodies have been studied in the greatest detail.

3. Equation (15.5) also gives more general forms of (15.1)–(15.3), and of the analogous equation for 6-connected systems, which is

$$0f_3 - 4f_4 - 8f_5 - 12f_6 - \cdots = 24$$

For example,

$$p = 3 \qquad 3f_3 + 2f_4 + f_5 \pm 0f_6 - f_7 - 2f_8 - \cdots = 6(2 - t) \quad (15.1a)$$

the solutions of which correspond to the (3-connected) duals of the triangulated polyhedra. For example, a 3-connected surface tessellation of 8-gons has $3(t - 2)$ 8-gons in the repeat unit. The solutions of (15.1a) are described under 3-connected nets.

4. There are also solutions of equations such as (15.1a) corresponding to 3D polyhedra with faces of more than one kind and all vertices similarly connected, which are the 3D homologues of the Archimedean solids, and their duals (with all faces n-gons but vertices of more than one kind), which are the homologues of the Catalan solids. We now consider briefly these "semiregular" 3D polyhedra.

Examples include (*a*) the polyhedron formed from truncated tetrahedra linked through octahedra to form a diamondlike structure having three triangular and two 6-gon faces meeting at each vertex; (*b*) the diamondlike arrangement of truncated octahedra linked through hexagonal prisms to four other truncated octahedra (the basic net of faujasite), with three 4-gons and one 6-gon meeting at each vertex; (*c*) the 6-tunnel polyhedron formed from truncated octahedra linked through cubes (the net of zeolite A), with two 4-gons and two 6-gons meeting at each vertex.

Like the triangulated bodies, these semiregular 3D polyhedra represent an extension of tessellations beyond those possible on the Euclidean plane. Systems of 4- and/or 8-gons include 4^3 (cube) and $4^2 8$ (octagonal prism), 4^4 and 4.8^2 (plane nets), and the higher members

$p =$	3	4	5
	4^3	4^4	4^5
	$4^2 8$	$4^3 8$	
	4.8^2	$4^2 8^2$	
	8^3		

(to the right of the heavy line) that are only realizable as 3D polyhedra. These symbols and the descriptions in the previous paragraph include only the polygons of the surface tessellations and if $p > 3$ they are not the complete point symbols of the p-connected net (see Chapters 2, 9, and 10).

Polyhedral space-fillings involving truncated cuboctahedra (*cot*) provide examples of 3D polyhedra which are homologues of the finite Archimedean solids. Polyhedra 4^26^2, 4^36, and 4^28^2 are included in Table 18.1 and are described in Chapter 18 as examples of complementary polyhedra.

Other semiregular 3D polyhedra have all faces of the same kind (e.g., triangles) but vertices of more than one kind—the homologues of the Catalan solids. For example, the polyhedron formed from icosahedra joined through four icosahedra to form a diamondlike structure has equal numbers of 5- and 8-connected vertices, 12 of each in the repeat unit. The relevant equation for triangulated polyhedra is

$$3c_3 + 2c_4 + c_5 \pm 0c_6 - c_7 - 2c_8 - \cdots = 6(2 - t)$$

which, like (15.1a), is derived from the modified Euler relation. Evidently the choice of a particular combination of vertex types does not uniquely define the polyhedron. For example, solutions for 4-tunnel polyhedra having c_5 and c_8 vertices include

c_5	12	36	60
c_8	12	24	36 etc.

which represent 3D polyhedra having icosahedra at the points of the diamond net joined through tunnels consisting, respectively, of 1, 3, 5, and so on, icosahedra that form linear tunnels by sharing a pair of opposite faces. Trivial solutions of any desired degree of complexity can be devised. The dual tessellations corresponding to these triangulated polyhedra form further families of 3-connected nets including, for example, the polyhedron consisting of equal numbers of 5- and 8-gons—compare the plane net with equal numbers of 4- and 8-gons.

16

The systematic derivation of three-dimensional polyhedra

For any specified number of tunnels we can calculate from the general equation (15.5) the values of Z for tessellations having p n-gons meeting at each point as, for example, Table 16.1 for $t = 6$. We are interested in the positive integral values of Z. These values decrease along a horizontal row or down a vertical column, and clearly there must be a lower limit for Z. Although one polyhedron has been found with $Z < t$, this seems to be a very special case [(3, 12) – $8t$, $Z = 6$], and on the somewhat arbitrary assumption that there will probably be no other cases with $Z < t$, we shall suppose that our study is limited to the (positive, integral) solutions lying between the heavy lines in Table 16.1; exceptions are (6, 5), (8, 4) and (3, 12). These solutions are excluded because their duals have $Z < t$, for a polyhedron is not realizable if its dual on the same surface cannot exist.

It is found that certain 3D polyhedra can be constructed from enantiomorphic units only if alternate units are D and L. The repeat unit is then strictly (D + L) and the value of Z is twice the value for a single enantiomorphic unit. This also follows from (15.5), for two adjacent t-tunnel units can be regarded as one $(2t - 2)$-tunnel unit, and for a given type of tessellation the value of Z for a unit with $2t - 2$ tunnels is twice that for a unit with t tunnels.

The following account is essentially descriptive. We give examples of polyhedra with $t = 3$, 4, 5, 6, 8, and 12; we have derived no examples of surfaces with t exceeding 12, which is the coordination number for the

Table 16.1 Values of Z for polyhedra with 6-tunnel repeat units

		p									
n	Z	3	4	5	6	7	8	9	10	11	12
3	$24/(p-6)$	−8	−12	−24	∞	24	12	8	6	*	4
4	$32/(2p-8)$	−16	∞	16	8	*	4	*			
5	$40/(3p-10)$	−40	20	8	5						
6	$48/(4p-12)$	∞	12	6							
7	$56/(5p-14)$	56	*								
8	$64/(6p-16)$	32	8								
9	$72/(7p-18)$	24	*								
10	$80/(8p-20)$	20	*								
11	$88/(9p-22)$										
12	$96/(10p-24)$	16									

* Z nonintegral.

closest packing of equal spheres. We do not give additional tables like Table 16.1 because it is more convenient to deal with these polyhedra as families $\{n, p\}$ with various values of t. All the polyhedra of Table 16.1 and analogous tables are not necessarily realizable as *nets* having no polygons smaller than n-gons, since this condition was not introduced in deriving the formula for Z.

For each family $\{n, p\}$ we set out the values of Z for the different values of t. In each case symbols are used to indicate that the value of Z is non-integral or that there is no solution having having $Z > t$. A number of the polyhedra are illustrated, and the figures show either a repeat unit or a larger portion of the infinite periodic polyhedron. To accentuate the surfaces (and to facilitate the making of the models), the models for n up to 6 can be constructed of strips of card joined together by paper fasteners. The edges of the n-gons would be the median lines of the strips and the p-connected points of the tessellation correspond to the centers of the rings of p paper fasteners. The duals of some of these polyhedra are illustrated as uniform 3D nets, (7, 3), (8, 3), and (9, 3), in Chapter 5. Being built with *straight* links (of equal length), these wire models are not strictly the duals of the corresponding $\{3, p\}$ polyhedra, for the true duals (e.g., Figs. 19.1 and 19.2) are tessellations of p-gons on the surface (of varying curvature) of the original polyhedron.

The $\{3, p\}$ family

This family represents the continuation into three dimensions of the series starting with (3, 3), tetrahedron, (3, 4), octahedron, (3, 5), icosahedron, and (3, 6), plane triangular net.

(3, 7)	t:	3	4	5	6	8	12
	Z:	6	12	18	24	36	60

Examples are given of the 3-, 4-, 6-, and 8-tunnel polyhedra.

Three coplanar tunnels ($Z = 6$)

The unit of Fig. 16.1*a* is enantiomorphic. Units of the same kind (D or L) join together to form the plane net (6, 3) (Fig. 16.2), the simplest 2D 3-

Fig. 16.1. Repeat units of (3, 7) polyhedra.

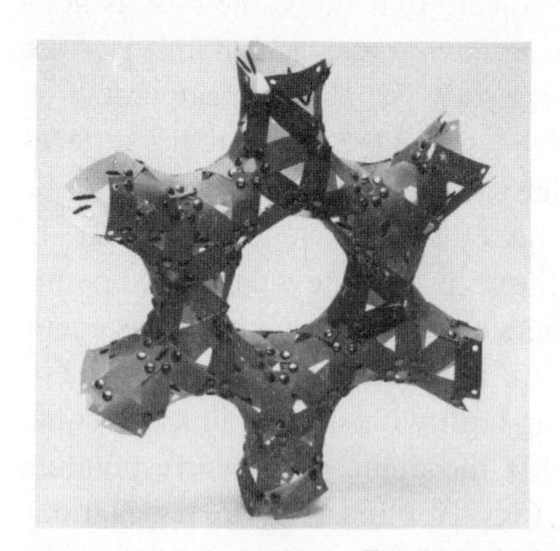

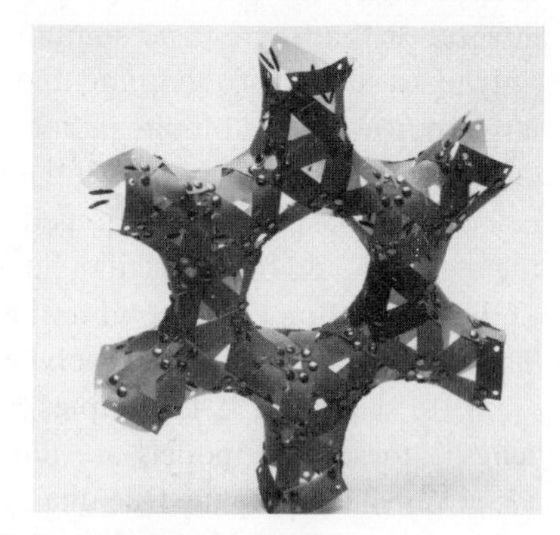

Fig. 16.2. A planar (3, 7) polyhedron.

connected net. [In the unit cell of the polyhedron there are 12 points, since there are 2 points in the unit cell of (6, 3).] By combining D and L units, 3D systems arise, of which the simplest is based on the (10, 3)-*b* net of Fig. 15.1*b*. This arises from rows of D units as shown in Fig. 16.3, which are joined at the points *a* to similar rows of L units in planes perpendicular to that of the paper. This polyhedron is illustrated in Fig. 16.4. In a net built of D and L units, the repeat unit is strictly the combination (D + L); that is, it is a 4-tunnel unit having in the present case $Z = 12$. [In this connection see also the (4, 7) 8-tunnel polyhedron.]

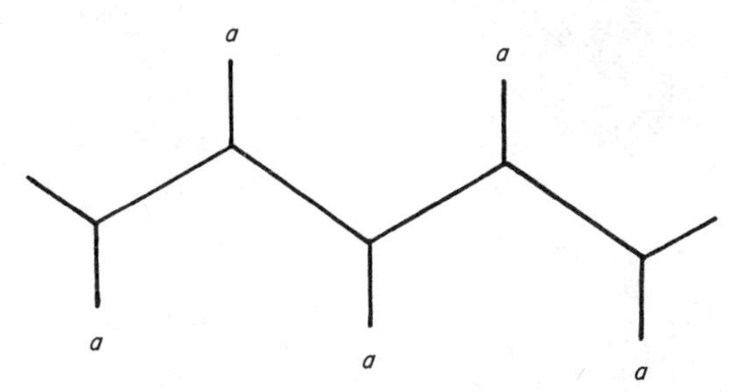

Fig. 16.3. Portion of the net (10, 3)-*b*.

Fig. 16.4. Polyhedron (3, 7) based on the net (10, 3)-*b*.

It was stated in Part 7 that the second (3, 7) 3-tunnel unit (Fig. 16.1*b*) forms a 3D polyhedron based on a distorted version of the net (10, 3)-*a*. This is not confirmed on reexamination of this unit, which is topologically equivalent to a trigonal prism with square antiprism tunnels. However the unit may be used to form polyhedra based either on the plane 6-gon net or on the net (10, 3)-*b*, which contain both 6- and 7-connected points. These polyhedra are equivalent to open packings of trigonal prisms connected through pairs (or

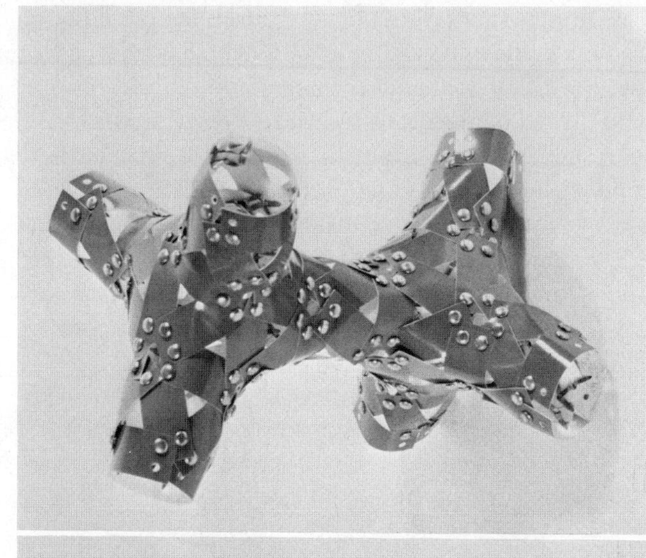

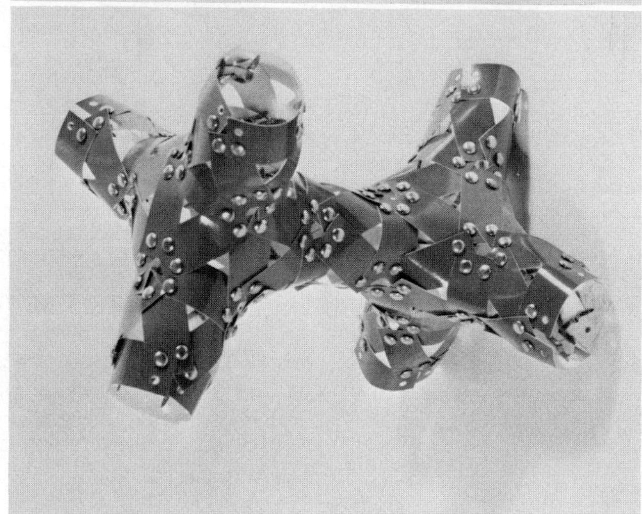

Fig. 16.5. Two units (D and L) of a tetrahedral (3, 7)-4*t* polyhedron.

any even number) of square antiprisms. The numbers of triangles and points would be as follows:

Number of antiprisms in each tunnel	Number of triangles in repeat unit	Points in repeat unit	
		c_6	c_7
0	2 + 12 = 14	0	6
2	24 = 26	6	6
4	48 = 50	18	6
etc.	etc.	etc.	

Their duals are 3-connected tessellations of 6-gons and 7-gons representing the transition from the plane 6-gon net (which can also be inscribed on an open cylinder) to the tessellations consisting entirely of 7-gons, which require a 3D polyhedral surface. Compare the mixed c_6–c_7 tessellations derived from the 6-tunnel (3, 7) unit and the c_6–c_8 tessellations derived from the 4-tunnel (3, 8) unit, noted shortly.

Four tunnels arranged tetrahedrally (Z = 12)

The two units (D and L) of a tetrahedral (3, 7)-4*t* polyhedron of Fig. 16.5 are related by a center of symmetry at the central point of the vertical tunnel. Such pairs build a 3D polyhedron based on the diamond net. This polyhedron consists of icosahedra at the points of that net joined through octahedral tunnels, and it is one of those that can be built with plane equilateral triangular faces. It is illustrated later as the regular triangulated 3D polyhedron I_{4t} (3, 7) (Fig. 17.10). A polyhedron based on the diamond net can be built from regular icosahedral units because four faces of an icosahedron may be selected such that the six angles between their normals all have the value $109°28'$. On this point see also the discussion of the second (3, 8)-4*t* unit on page 217.

Six tunnels arranged octahedrally (Z = 24)

The nucleus of the 6-tunnel unit of Fig. 16.1*c* is a snub cube, which is enantiomorphic. These units may be joined together to form a 3D (3, 7) polyhedron based on the *P* lattice, when the tunnels correspond to square antiprisms, but a slight twist is required at each junction because the relation between opposite square faces of a snub cube is not the same as that between the opposite faces of a square antiprism. The same nucleus may be used to form

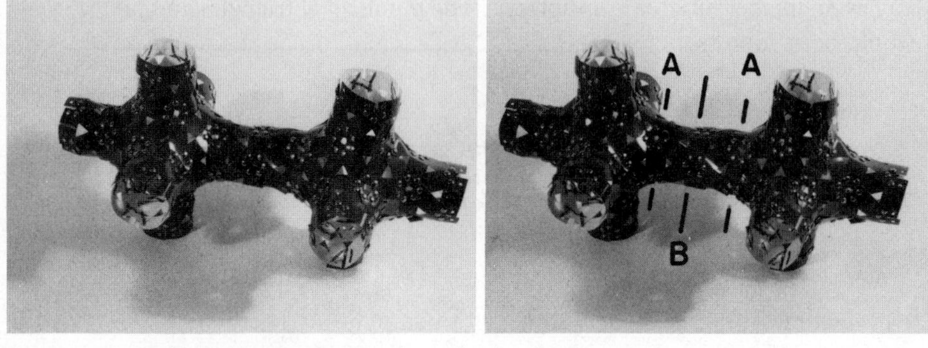

Fig. 16.6. Formation of $\left(3, \begin{matrix}6\\7\end{matrix}\right)$ or (3, 8) polyhedra (see text).

other 3D polyhedra. If reflected across the bases of the tunnels (square faces of the snub cubes) so that the planes *A* in Fig. 16.6 are mirror planes, there is formed a (3, 8) polyhedron based on D and L snub cubes. This is the complement of the polyhedron formed from the (3, 8)-6*t* unit of Fig. 16.8*b*; both the polyhedron and its complement are illustrated later as regular 3D polyhedra I_{6t} (3, 8). Alternatively, if this unit is reflected across the plane *B* in Fig. 16.6, the points in that plane become 6-connected, and the 3D polyhedron has units of 24 c_7 at its nodes joined by cylindrical tunnels around which portions of the triangulated c_6 net are inscribed. This mixed $\left(3, \begin{matrix}6\\7\end{matrix}\right)$ polyhedron may be constructed with $(4 + 8m)$ 6-connected points on each tunnel, corresponding to D and L snub cubes at alternate points of the *P* lattice joined through tunnels consisting of even numbers of square antiprisms. Assuming all tunnels to be equal in length, so that the 3D polyhedron is based on the cubic *P* lattice, we have a family of 3D polyhedra:

c_7:	24	24	24	24
c_6:	0	12	36	60 etc.

that can be compared with the icosahedron and the related polyhedra having additional 6-connected vertices, $12c_5 + n\,c_6$.

Eight tunnels ($Z = 36$)

An 8-tunnel unit is illustrated in Fig. 16.7. This unit does not possess cubic symmetry (only *mmm*); accordingly there is no complementary polyhedron

Fig. 16.7. Repeat unit of (3, 7)-8*t*.

based on identical units with four coplanar tunnels (see p. 251). The dual is the uniform 3-connected net (7, 3)-*d*.

(3, 8)	t:	3	4	5	6	8	12
	Z:	3	6	9	12	18	30

Examples are given of units of all these types.

Three coplanar tunnels **(*Z* = *3*)**

The 3-tunnel unit of Fig. 16.8*a* builds a 3D polyhedron based on the net (10, 3)-*b*. This polyhedron has $y = 2$. The dual is the uniform net (8, 3)-*e*. Another 3-tunnel unit gives the polyhedron of Fig. 16.9 based on the net (10, 3)-*a*. It is equivalent to octahedra joined through tetrahedral tunnels, and when constructed with plane equilateral triangular faces, it is the regular

Fig. 16.8. Repeat units of (3, 8) polyhedra.

Fig. 16.9. A (3, 8) polyhedron based on the net (10, 3)-*a*.

3D polyhedron O_{3t} (3, 8). In this polyhedron there are links with $y = 2, 3$, and 4, with weighted mean $2\frac{3}{4}$. One ring of the dual (8, 3) polyhedron (net) is illustrated in Fig. 5.34.

***Four tunnels arranged tetrahedrally* ($Z = 6$)**

Two tetrahedral units have been constructed. The first builds the polyhedron of Fig. 16.10 based on the diamond net. There are 3-gons around the tunnels (in addition to those on the surface) with the result that for half the links $y = 3$; for the remainder $y = 2$, giving $y_{\text{mean}} = 2\frac{1}{2}$. The nucleus of this unit is an octahedral group of six points, and the 3D polyhedron is equivalent to octahedra at the points of the diamond net that share four faces (arranged tetrahedrally) with a second set of octahedra that form the tunnels. The latter octahedra share a pair of opposite faces with the former. This poly-

Fig. 16.10. A (3, 8) polyhedron based on the cubic diamond net.

Fig. 16.11. A second (3, 8)-4*t* unit.

hedron can be constructed with plane equilateral triangular faces and is illustrated as the regular 3D polyhedron O_{4t} (3, 8) in Fig. 17.7. The dual has 6-gon in addition to 8-gon circuits and is therefore not an (8, 3) net.

The second unit (Fig. 16.11) also builds, with some distortion, a diamond-like polyhedron for which $y_{\text{mean}} = 2\frac{1}{4}$. Here also there are 3-gons around the tunnels, so that for one-quarter of the links $y = 3$. A portion of the dual (8, 3) net is illustrated in Fig. 5.35. The reason for this unit's inability to build a *cubic* diamondlike polyhedron raises a point of more general interest. The nucleus of the unit is an icosahedron, and the 3D polyhedron is built from icosahedra, each sharing four faces arranged tetrahedrally. [This is the same unit as in the (3, 7)-4*t* polyhedron in which the icosahedra are joined through octahedral tunnels.] An icosahedron with four such faces differentiated from the remainder (e.g., by coloring) is enantiomorphic. Four faces can be selected such that the angles between their normals are 109°28′, with the result that the four tunnels from any unit are directed in this regular way. However the relative spatial orientation of the tunnels from adjacent units having a face in common [as opposed to being joined through octahedral tunnels as in (3, 7)-4*t*] is not that required for the cubic diamond structure, that is, exactly the staggered (*trans*) relationship. In an icosahedron resting on the central face (broken lines) of Fig. 16.12, the directions of the tunnels are represented by lines *D*, which are the normals to the three shaded faces, and the fourth tunnel is perpendicular to the paper and is directed upward from the center of the icosahedron. If the upper (heavily drawn) face is shared with another icosahedron of the same chirality, the axes of its tunnels (pointing upward) project as the lines *U*. These are not in the staggered relation to the lines *D*, as is required to form a *cubic* diamond net. (If the

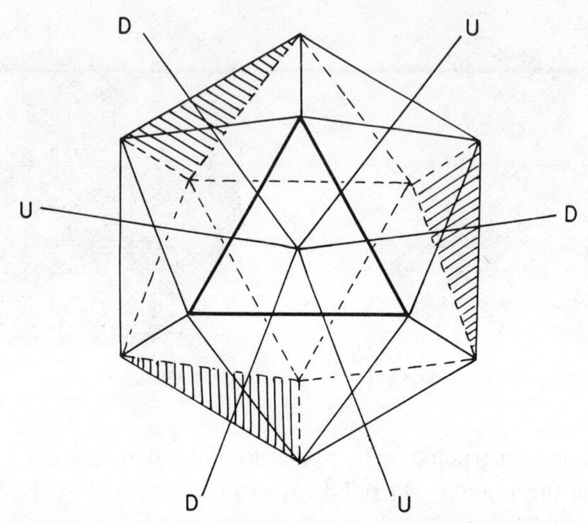

Fig. 16.12. Relation between face-normals of two icosahedra that share a face.

icosahedra were alternately of opposite chirality, the relative arrangement of tunnels from adjacent units would be eclipsed—*cis*.) Thus failure to form a 3D polyhedron is associated with the orientation of the *edges* of the shared face relative to the tunnel directions. This problem is not peculiar to this (3, 8)-4*t* unit; it explains, for example, why an 8-tunnel polyhedron (8, 4) cannot be formed from truncated cubes sharing all their triangular faces and why the (3, 7)-6*t* unit formed from snub cubes and antiprisms is distorted.

***Four coplanar tunnels* (*Z* = *6*)**

The simplest possibility here is the formation of the plane (4, 4) net. This requires that all the tunnels from adjacent units lie in the same plane (Fig. 16.13*a*). If adjacent repeat units are related by a rotation through 90°, every

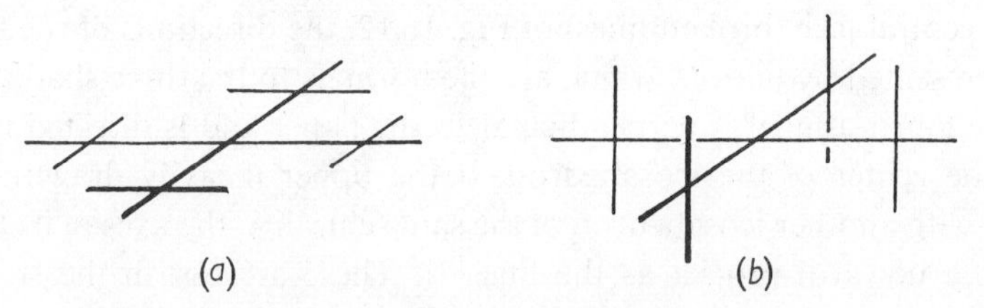

Fig. 16.13. Relation of adjacent units in 4-tunnel systems.

point is surrounded as in Fig. 16.13*b*. The 3D polyhedron then has the form of the NbO net.

The unit of Fig. 16.14*a* does not join together to form a polyhedron based on the planar 4-gon net but joins as in Fig. 16.13*b* to form a 3D polyhedron based on the NbO net (Fig. 16.15). This unit, however, can form a polyhedron based on the plane 4-gon net if it is reflected across the plane indicated in Fig. 16.14*b*. The polyhedron so formed is not a (3, 8) but a $\left(3, {6 \atop 8}\right)$, having an even number of 6-connected points along a tunnel and, of course, a group of six 8-connected points around each node of the net. A portion of the simplest tesselation (polyhedron) of this family, $6c_8 + (4m)c_6$, appears in Fig. 16.14*b*.

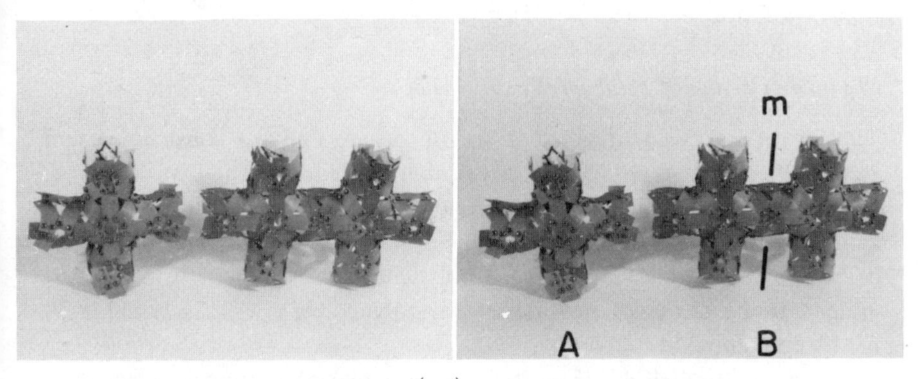

Fig. 16.14. Formation of $\left(3, {6 \atop 8}\right)$ or (3, 8) polyhedron (see text).

Fig. 16.15. Polyhedron (3, 8) based on the NbO net (Fig. 15.1d).

Since the (body-centered) NbO net has six points in the unit cell, the polyhedron of Fig. 16.15 contains 36 8-connected points. Around the tunnels the shortest circuits of points form 4-gons; therefore $y = 2$. The shortest circuit of triangles around the tunnels is one of eight triangles. The dual (Fig. 5.36) is therefore a uniform (8, 3) net in which there are no circuits smaller than 8-gons.

Five trigonal bipyramidal tunnels ($Z = 9$)

The five tunnels are not equivalent; there are three in one plane linking the units into a (6, 3) net and two perpendicular to this plane linking together the sheets to form the 5-connected net of Fig. 15.1*e*. The polyhedron is illustrated in Fig. 16.16. For all except three of the 36 links in the repeat unit, $y = 2$ (for the others $y = 3$), giving $y_{\text{mean}} = \frac{25}{12}$.

Six tunnels arranged octahedrally ($Z = 12$)

The unit of Fig. 16.8*b*, in which all six tunnels are similarly constructed, can be placed at the nodes of the primitive cubic lattice to form a 3D polyhedron, which is illustrated in its most symmetrical form (with plane equilateral faces) as the regular polyhedron I_{6t} (3, 8) in Fig. 17.11. The complementary polyhedron, which is different from the polyhedron itself, has been mentioned under 6-tunnel (3, 7) polyhedra. For this polyhedron $y = 2$. The dual (8, 3) net is represented in Fig. 5.39.

A second 6-tunnel unit (Fig. 16.8*c*) has the two polar tunnels larger than the equatorial ones. This polyhedron is identical to its complement; each fills one-half of space. The dual (8, 3) net is illustrated in Fig. 5.40.

Eight hexagonal bipyramidal tunnels ($Z = 18$)

The polyhedron complementary to the 5-tunnel one of Fig. 16.16 has six equatorial tunnels and two polar tunnels, and the latter differ in structure from the equatorial ones.

Eight cubic tunnels ($Z = 18$)

The polyhedron complementary to the 4-tunnel polyhedron of Fig. 16.15 is built from 8-tunnel units; the x and y values have already been given. A second 8-tunnel unit is illustrated in Fig. 16.17, and this also builds a polyhedron based on the b.c.c. lattice. Of the 72 links in the repeat unit, 52 have

Fig. 16.16. A polyhedron (3, 8) based on the 5-connected net of Fig. 15.1*e*.

Fig. 16.17. A (3, 8)-8*t* repeat unit.

$y = 2$, 16 have $y = 3$, and 4 have $y = 4$, giving $y_{\text{mean}} = 2\frac{1}{3}$. The dual (8, 3) net is illustrated in Fig. 5.41.

Twelve tunnels ($Z = 30$)

A 12-tunnel unit was illustrated in Part 7. Further work is required to check that this unit builds a 12-tunnel polyhedron.

(3, 9)	t:	3	4	5	6	8	12
	Z:	*	4	6	8	12	20

* No solution having $Z > t$.

Models of three (3, 9) polyhedra have been constructed from 4-, 6-, and 8-tunnel units. The 4-tunnel unit (Fig. 16.18) can be built with plane equilateral faces; it is described with the other regular 3D polyhedra [as T_{4t} (3, 9)] and is illustrated in Fig. 17.4.

A 6-tunnel unit is shown in Fig. 16.19*a*. The dual net, (9, 3)-*b*, is illustrated in Fig. 5.22.

The 8-tunnel unit of Fig. 16.20*a* forms a 3D polyhedron based on the b.c.c. lattice. Like the 4-tunnel polyhedron, it may be constructed with plane

Fig. 16.18. Repeat unit of (3, 9)-4*t*. The same polyhedron built with plane equilateral faces is T_{4t} (Fig. 17.4).

Fig. 16.19. (*a*) A (3, 9)-6*t* repeat unit. (*b*) A (3, 10)-6*t* repeat unit.

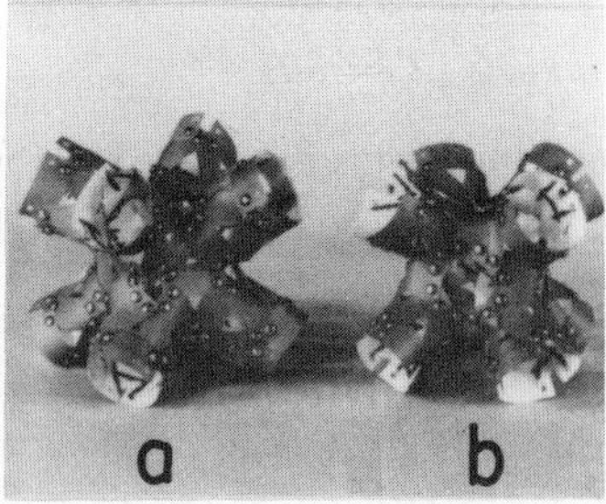

Fig. 16.20. Repeat units of (*a*) (3, 9)-8*t* and (*b*) (3, 12)-8*t*. The same polyhedra built with plane equilateral faces are I_{8t} (Fig. 17.13) and O_{8t} (Fig. 17.9).

equilateral faces; it is illustrated as the regular polyhedron I_{8t} (3, 9) in Fig. 17.13. The complementary polyhedron, with four coplanar tunnels from each repeat unit, is illustrated as a regular 3D polyhedron in Fig. 17.14.

(3, 10)	t:	3	4	5	6	8	12
	Z:	*†	*	*†	6	9	15

A 6-tunnel unit is illustrated in Fig. 16.19b, and a unit cell of the plane-faced 3D polyhedron in Fig. 16.21. Regarding the latter as a 10-connected net—that is, taking account of the 3-gon circuits around the tunnels—we find that all the points are equivalent, with $x = 16$. The links are of two kinds; four-fifths have $y = 3$, and the remainder (the short edges of the isosceles triangles that enclose the tunnels in Fig. 5.8) have $y = 4$, giving $y_{\text{mean}} = \frac{16}{5}$. This polyhedron is described in more detail, together with the dual net (10, 3)-g on page 40. For another (3, 10)-6t polyhedron, see Fig. 17.8.

Fig. 16.21. Unit cell of (3, 10)-6t built with plane faces. The (curved-surface) repeat unit is shown in Fig. 16.19b.

(3, 11) Calculation of Z for these polyhedra shows that Z is integral only for $t = 7 + 5m$; for example, $Z = 6$ for $t = 7$ and $Z = 12$ for $t = 12$. It seems unlikely that any of these systems can be realized.

(3, 12) For all these polyhedra $Z = t - 2$. The 8-tunnel polyhedron ($Z = 6$) based on the unit of Fig. 16.20b is the only known example of a

* No solution having $Z > t$.
† Z nonintegral.

polyhedron with $Z < t$. The regular form of this polyhedron with plane equilateral faces is illustrated in Fig. 17.9 as a regular 3D polyhedron.

The $\{4, p\}$ family

This family starts with the cube (4, 3), which is followed by the plane net (4, 4). All higher members are polyhedra based on 2D or 3D nets.

(4, 5)	t:	3	4	5	6	8	12
	Z:	4	8	12	16	24	40

Nine polyhedra of this class are now known, if we count as different polyhedra the three pairs of complementary polyhedra based on the two different 4-tunnel units and one 5-tunnel unit of Table 16.2, namely, types *a* and *h*, *b* and *i*, and *c* and *g*.

Each of the three 6-tunnel polyhedra is identical to its complement. Because we deal later with the (5, 4) class, whose members are the duals of the (4, 5) polyhedra, and because all except one of the latter can be constructed with (plane) square faces and are therefore illustrated in the section on regular 3D polyhedra, we limit the illustration here to four of the polyhedra with curved faces (Table 16.2).

Table 16.2 Key to illustrations of (4, 5) and (5, 4) polyhedra

		Figure identifying		
		(4, 5)		(5, 4)
Type	Polyhedron	Curved faces	Plane faces	Curved faces
a	4 coplanar tunnels	16.22*a*	17.15	—
b	4 coplanar tunnels	16.22*b*	17.16	16.24*a*
c	5 trigonal bipyramidal tunnels	—	17.18	16.25*a*
d	6 tunnels	16.22*c*	—	16.25*b*
e	6 tunnels	—	17.19	—
f	6 tunnels	—	17.19	—
g	8 tunnels, complement of *c*	—	—	—
h	8 (equivalent) tunnels complement of *a*	—	—	16.24*b*
i	8 (equivalent) tunnels complement of *b*	16.22*d*	17.20	—

Fig. 16.22. (*a*), (*b*) Pairs of repeat units of two polyhedra (4, 5)-4*t*; (*c*), (*d*) Repeat units of (4, 5)-6*t* and (4, 5)-8*t*.

One 4-tunnel unit ($Z = 8$) (Figure 16.22*a*) can be joined to similar units to form either a plane or 3D system. In the former case the basic net is the plane (4, 4) net, whereas in the 3D case adjacent units are related by a rotation through 90°, and the basic net is the 4-connected NbO net. This polyhedron and its complement are realizable with plane square faces and are included with the regular 3D polyhedra (Fig. 17.15). A second 4-tunnel unit (Fig. 16.22*b*) also forms a polyhedron based on the NbO net, a larger portion of which is shown in Fig. 19.1, and the regular configuration in Fig. 17.16. A unit of the complementary (curved surface) polyhedron appears in Fig. 16.22*d* and a model of the regular polyhedron in Fig. 17.20.

The 6-tunnel unit that forms a polyhedron based on the *P* lattice (Fig. 16.22*c*) is of special interest as the only polyhedron of this group that cannot be constructed with plane square faces.

(4, 6)							
	t:	3	4	5	6	8	12
	Z:	*	4	6	8	12	20

It is interesting that the values of Z for this class are the numbers of vertices of the regular solids, as in the case of (3, 9).

Two 6-tunnel units are illustrated in Fig. 16.23; the links of Fig. 16.23*a*, which is Coxeter's $\{4, 6/4\}$, correspond to the primitive lattice.

(4, 7)							
	t:	3	4	5	6	8	12
	Z:	*†	*†	†	*†	8	†

* No solution having $Z > t$.
† Z nonintegral.

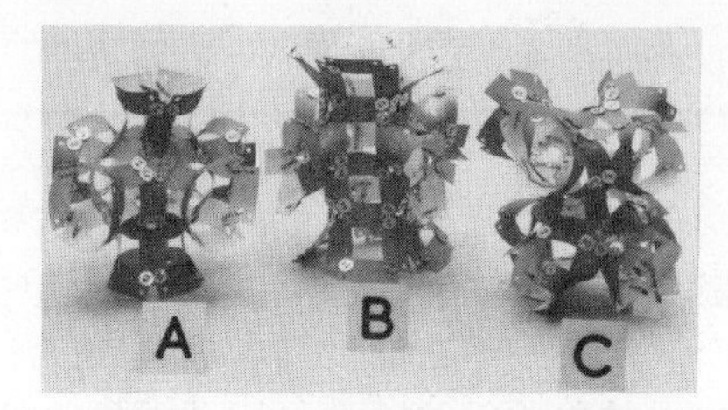

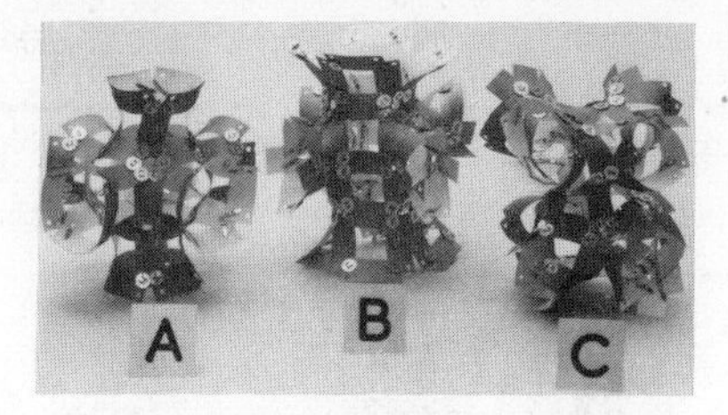

Fig. 16.23. (*a*), (*b*) Two (4, 6)-6t repeat units; (*c*) a (4, 7)-8t unit.

There is only one solution in this, the last class of the (4, p) family. Figure 16.23c illustrates one form of the 8-tunnel unit; the top half is the mirror image of the lower half.

In (4, 8) $Z = t - 2$ throughout; no examples have been found.

The $\{5, p\}$ family

The first member is the pentagonal dodecahedron (5, 3). There is no plane net with an integral value of p; instead there is the net $\left(5, \frac{3}{4}\right)$. The classes from (5, 4) onward are 3D polyhedra. We list three classes in this family, though it is improbable that any polyhedra (5, 6) could be realized.

(5, 4)	t:	3	4	5	6	8	12
	Z:	5	10	15	20	30	50

Members of this class are the duals of the (4, 5) polyhedra listed in Table 16.2, and only four are illustrated. Figure 16.24a shows a pair of 4-tunnel units of

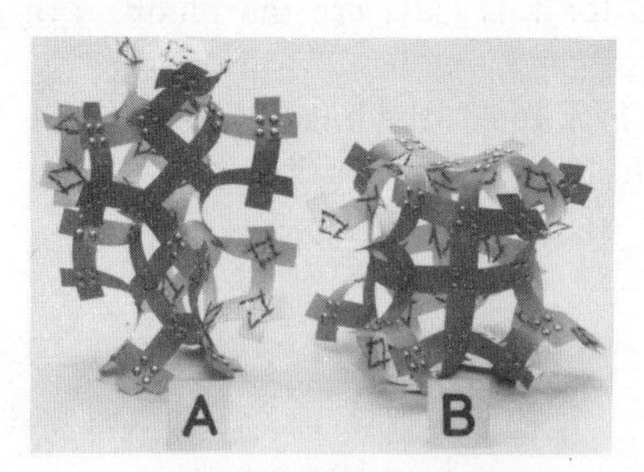

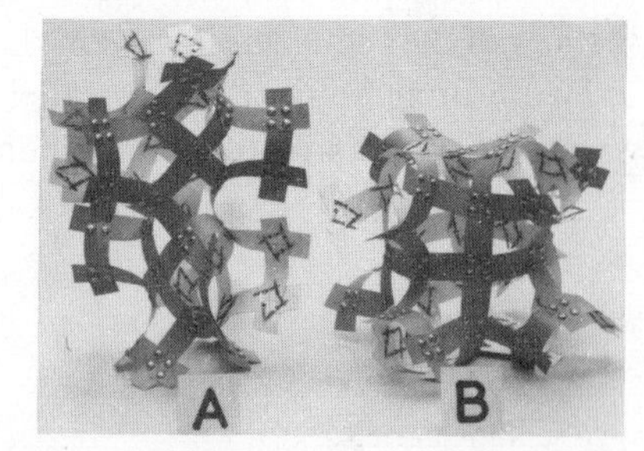

Fig. 16.24. (*a*) Pair of (5, 4)-4t units; (*b*) a (5, 4)-8t unit.

type b; a larger portion of this polyhedron appears in Fig. 19.2. Polyhedra of types c and d of Table 16.2 are given in Fig. 16.25*a*, and *b*, and an 8-tunnel unit of type h is in Fig. 16.24*b*.

(5, 5)	t:	3	4	5	6	8	12
	Z:	*	4	6	8	12	20

This class has the same numerical values of Z as (3, 9) and (4, 6); it is of special interest as the continuation of the series (3, 3), tetrahedron, and (4, 4), plane net. The dual of any polyhedron in this class is identical with the original polyhedron.

The 4-tunnel ($Z = 4$) unit of Fig. 16.26*a* is remarkable for possessing a unique $\bar{4}$ axis, which is somewhat unexpected for a system (5, 5). There are two ways of making each junction to neighboring units, only one of which leads to a 3D polyhedron. If all junctions are made so that the $\bar{4}$ axes of all units are parallel, a 3D polyhedron based on the "compressed-diamond" net is formed (see p. 117). There are no circuits smaller than 5-gons; therefore the 3D polyhedron is also a (5, 5) net. This net can be constructed with straight links and a trigonal bipyramidal arrangement of links around each point, when the axial links have to be made somewhat longer than the equatorial ones (Fig. 16.27). This net is uniform if we accept the "realistic" viewpoint expressed on page 15, since all points are equivalent and have the point symbol 5^9 or, if we include the smallest circuit including a pair of axial links, $5^9 7$. The net has cubic symmetry, and the projections (Fig. 16.28) show that it may be derived from two interpenetrating (10, 3)-*a* nets of opposite chirality by introducing further links along the 3-fold axes; no points are added. It is therefore described by the position 16(b) in $Ia3d$ (cf. Fig. 11.1). The links of a 3D polyhedron based on a 4-tunnel unit ($Z = 4$) and a diamondlike net form a *body-centered* 3D net with $Z = 16$ because the underlying net of the polyhedron is the "compressed-diamond" net, which has a *cubic* unit cell containing four points. Figure 16.29 gives the positions of the centers of the 4-tunnel units of the 3D polyhedron in a projection of the cubic unit cell of (5, 5). This cell contains four such units ($Z = 4$), therefore contains 16 points. The broken lines indicate the cell of the "compressed-diamond" net on which the polyhedron (5, 5) is based. This cell has $c/a = 1/\sqrt{2}$ and contains eight 4-tunnel units ($Z = 32$). The unit cells of some special configurations of the diamond net are listed in Table 9.4.

* No solution having $Z > t$.

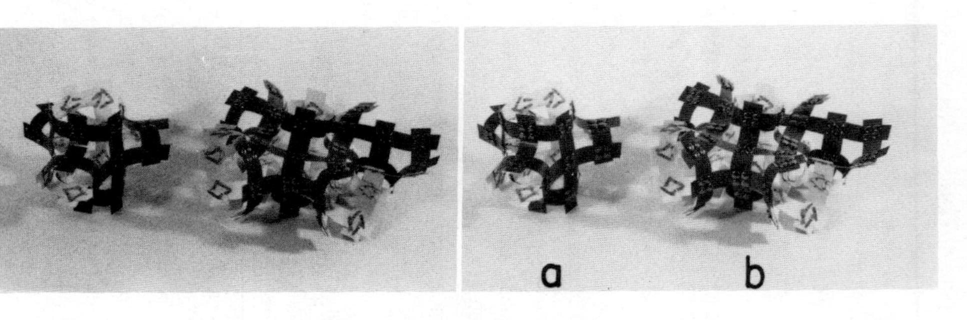

Fig. 16.25. Repeat units of (*a*) (5, 4)-5*t* and (*b*) (5, 4)-6*t*.

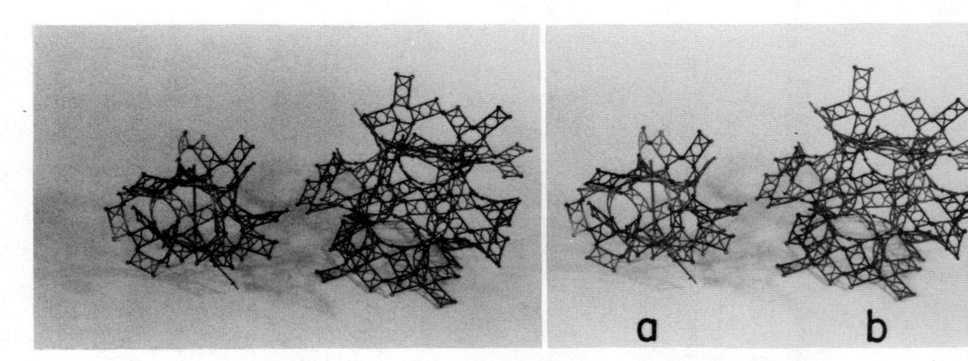

Fig. 16.26. Repeat unit of (*a*) (5, 5)-4*t* showing $\overline{4}$ axis; (*b*) (6, 4)-6*t*.

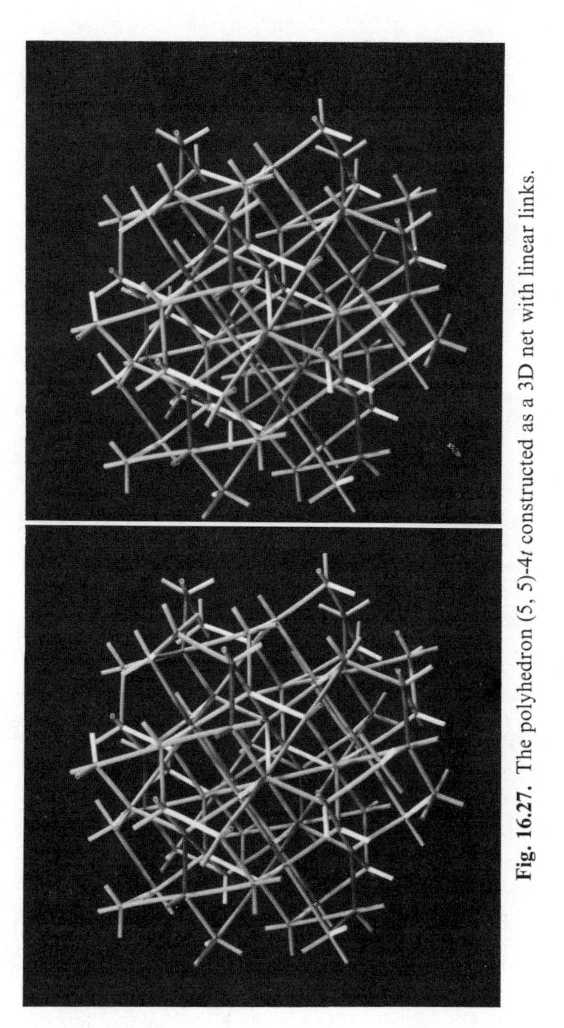

Fig. 16.27. The polyhedron (5, 5)-4*t* constructed as a 3D net with linear links.

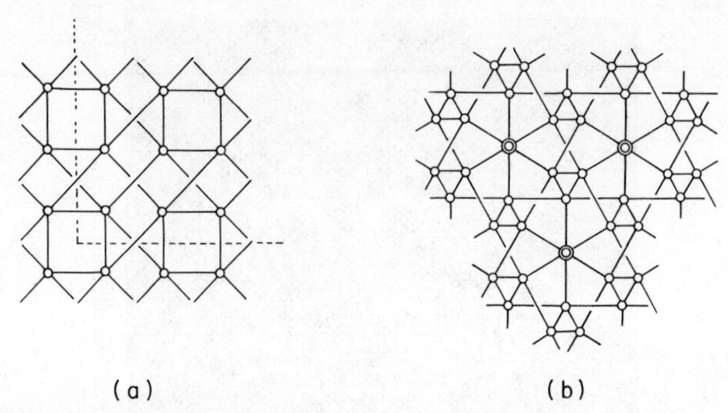

Fig. 16.28. The (5, 5) net of Fig. 16.27. Projections along (*a*) [100] and (*b*) [111].

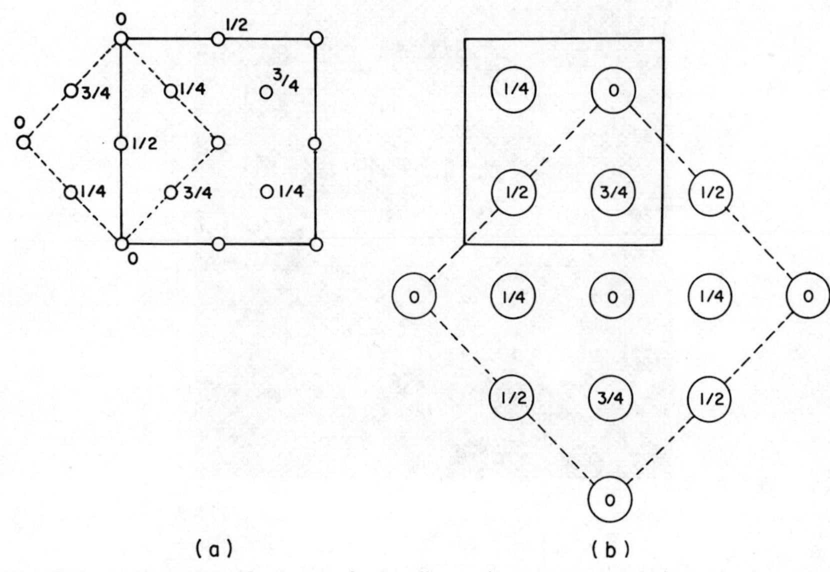

Fig. 16.29. (*a*) Projection of unit cell of cubic diamond net ($Z = 8$) and alternative b.c. tetragonal cell (dotted), $Z = 4$, $c/a = \sqrt{2}$. (*b*) Cubic unit cell of (5, 5) showing positions of centers of 4-tunnel units. The dotted lines indicate the cell of the "compressed diamond" net on which the 3D polyhedron (5, 5) is based.

The complementary polyhedron is identical with the polyhedron itself, and therefore this (curved-surface) polyhedron fills one-half of space. This is the only example of a 3D polyhedron (n, p) and complement with bivalent tunnels based on a pair of interpenetrating diamond nets (see p. 250).

This polyhedron (5, 5) continues the series

(3, 3)	(4, 4)	(5, 5)
Finite polyhedron	Plane net	3D polyhedron
$Z = 4$	$Z = 1$	$Z = 4$

In the class (5, 6) $Z \geqslant t$ only for $t = 10 + 4m$:

(5, 6)	t:	10	14	18
	Z:	10	15	20 etc.

No examples of polyhedra in this class have been found.

The {6, *p*} family

This family represents the continuation into three dimensions of the series starting with the plane net (6, 3). There are two classes with $Z \geqslant t$.

(6, 4)	t:	3	4	5	6	8	12
	Z:	3	6	9	12	18	30

The values of Z are the same as for (3, 8). Figure 16.26*b* shows the 6-tunnel unit that is the dual of the 6-tunnel (4, 6) of Figure 16.23*b*. This unit has tetragonal symmetry, having two large axial tunnels and four smaller equatorial ones. The polyhedron is identical with its complement. Another 6-tunnel unit, the dual of that of Fig. 16.23*a*, is the skew polyhedron {6, 4/4} of Coxeter. For these two polyhedra, regarded as 3D nets (see p. 14).

(6, 5)	t:	3	4	5	6	8	12
	Z:	*†	*	*	6	9	15

* No solution having $Z > t$.

† Z nonintegral.

The values of Z are the same as for (3, 10). No polyhedra of this class have been built.

The $\{7, p\}$ family

Polyhedra (or nets) (7, 3) are duals of the (3, 7) polyhedra we have already considered. In the class (7, 4) $Z \geqslant t$ only for $t = 5 + 3m$:

t:	3	4	5	6	8	12
Z:	*	*	7	*	14	*

The 8-tunnel polyhedron is the dual of the (4, 7) of Fig. 16.23*c*. Since it contains nonequivalent points and also 6-gons, it is not a uniform 7^6 net.

The $\{8, p\}$ family

Polyhedra (8, 3) are duals of (3, 8) polyhedra; those which are uniform (8, 3) nets are described in Chapter 5. The Z values are equal to $\frac{8}{3}$ times those of the corresponding (3, 8) polyhedra.

For (8, 4) the Z values are the same as for (4, 6) and equal to $2(t - 2)$; no examples have been found.

The $\{9, p\}$ family

Polyhedra (9, 3) are duals of (3, 9) polyhedra and have $Z = 6(t - 2)$. One that is a uniform (9, 3) net is illustrated in Fig. 5.22.

For (9, 4) the Z values are integral only for $t = 7 + 5m$; no examples have been found.

* Z nonintegral.

17

Regular three-dimensional polyhedra: The homologues of the Platonic solids

The infinite periodic 3D polyhedra that can be constructed with the same number of regular plane faces meeting at each vertex are called regular 3D polyhedra. They represent the continuation into three dimensions of the families starting with the five regular convex (Platonic) solids and the three

Table 17.1 Regular polyhedra and plane nets

Regular solids	Regular plane nets	Regular 3D polyhedra
(3, 3) (3, 4) (3, 5)	(3, 6)	(3, 7) (3, 8) (3, 9) (3, 10) (3, 12) (three)* (two)†
(4, 3)	(4, 4)	(4, 5) (4, 6) (five) (two)
(5, 3)	—	—
—	(6, 3)	(6, 4) (6, 6)‡

* These numbers do not include complements.
† A variant of one of these is also described.
‡ See text, under Regular Polyhedra $\{6, p\}$.

regular plane nets. They have not been studied exhaustively by the writer, but it seems worthwhile to summarize here what is at present known about these bodies, which are listed in Table 17.1.

Regular polyhedra $\{3, p\}$

The triangulated Platonic solids have the following numbers of vertices: 4 (tetrahedron), 6 (octahedron), and 12 (icosahedron). These finite polyhedra may be placed at the nodes of the basic cubic nets and joined through tunnels that must also be bounded exclusively by equilateral triangles. If there is to be the same number of faces meeting at each vertex of the 3D polyhedron, the tunnels must not contain vertices other than those which are to be shared with the finite polyhedra placed at the nodes of the net. Therefore the tunnel must be one of the following: (*a*) a tetrahedron sharing two faces with the polyhedra at adjacent nodes—these polyhedra necessarily have a common edge, which is an edge of the tetrahedron forming the tunnel; (*b*) an octahedron sharing *one* face at each end with the finite polyhedra; or (*c*) a "dodecahedral" tunnel that can link at each end through *two* faces of each polyhedron.

Tunnel *c* is closely related to the octahedron, from which it is derived by slitting along the edges accentuated in Fig. 17.1*a*, then simultaneously compressing along the directions *AB* and *CD*. If *AB* and *CD* are made equal in length to the other edges, this object represents a group of five regular tetrahedra, one on each face of a central tetrahedron (Fig. 17.1*b*), though we shall not use this very symmetrical form of this tunnel.

Only one polyhedron involving tetrahedral tunnels has been discovered. The octahedral tunnel shares one face of the central finite polyhedron and the dodecahedral tunnel two (adjacent) faces; thus having regard to the

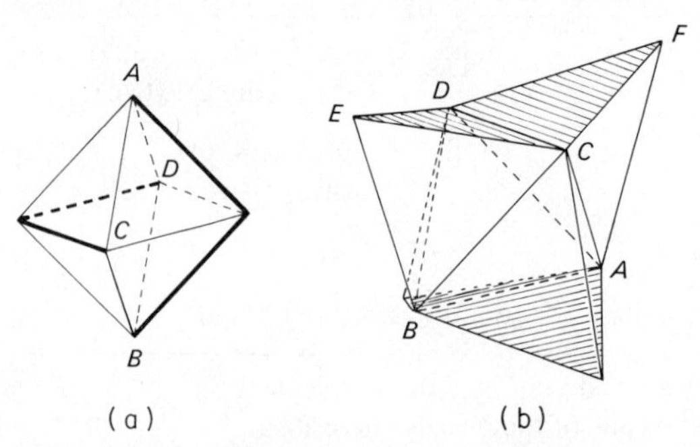

Fig. 17.1. Relation of dodecahedral tunnel to octahedron (see text).

symmetry of the central polyhedron, the systems to be studied are those set out in Table 17.2. All three polyhedra with octahedral tunnels can be constructed but only one of those with dodecahedral tunnels. The shape of this tunnel can be adjusted within certain limits, but it is not possible to link octahedra through either three or four of these tunnels or *regular* icosahedra through six such tunnels. The dihedral angle of a regular icosahedron is 138°12′, as compared with the value 148°24′ of the angle DCE/DCF for the tunnel built of regular tetrahedra. However a modified tunnel, with the eight exterior faces equilateral triangles and the shaded faces of Fig. 17.1*b* isosceles, fits over pairs of adjacent faces of a "cubic icosahedron." This solid, with vertices $(\frac{1}{2}\frac{1}{4}0)$, $(\frac{1}{2}\frac{3}{4}0)$, and so on, has eight equilateral faces (edge length $\frac{1}{2}\sqrt{6}$) and six pairs of isosceles faces with the common edge of unit length. (The dihedral angle over this edge is 126°52′.) It is also necessary to check whether further 3D polyhedra arise from Archimedean solids by sharing 4-gon, and so on, faces. The only possibility for triangulated polyhedra appears to be the system of snub cubes (alternately D and L) forming the polyhedron complementary to I_{6t}. Because this polyhedron can be so derived, it is evidently not essential to introduce the dodecahedral tunnel, since I_{6t} is the only polyhedron arising from its use. However the derivation of I_{6t} in this way from an icosahedron (albeit a nonregular "cubic" one) simplifies Table 17.2; the alternative is to derive I_{6t} from the snub cube.

Table 17.2 The triangulated regular 3D polyhedra

Polyhedron at nodes of net	Z	Tetrahedral tunnels	Octahedral tunnels	Dodecahedral tunnels	p	f_3
Octahedron	3	O_{3t}	—	—	8	8
Tetrahedron	4	—	T_{4t}*	—	9	12
Octahedron	6	—	O_{4t}	—	8	16
—	—	—	O_{6t}	(3)†	10	20‡
—	—	—	O_{8t}	(4)†	12	24‡
Icosahedron	12	—	I_{4t}	—	7	28
—	—	—	—	I_{6t}§	8	32‡
—	—	—	I_{8t}	—	9	36‡

* Also a variant T'_{4t} described in the text.
† Not realizable.
‡ See text.
§ Not a regular icosahedron.

We derive in the foregoing ways eight triangulated 3D regular polyhedra that form a complete series in which the number of faces (f_3) in the repeat unit increases in steps of four from 8 to 36. Of these, the three 4-tunnel ones are related in a very simple way to the corresponding finite regular solids.

	p						
	3	4	5	6	7	8	9
	Tetrahedron	Octahedron	Icosahedron	(Plane net)	I_{4t}	O_{4t}	T_{4t}
Z	4	6	12		12	6	4

A further 3D triangulated polyhedron (3, 9) is derived from T_{4t} (3, 9) by addition of further tunnels; it is described shortly as T'_{4t}. We noted in a footnote to Table 17.1 that the numbers of polyhedra (n, p) do not include complements. We are using the term complementary polyhedron here in the restricted sense noted on page 253; we therefore recognize only those of the 6- and 8-tunnel polyhedra marked with a double dagger (‡) in Table 17.2.

O_{3t}

The polyhedron of Fig. 17.2 is formed by placing octahedra at the nodes of the cubic net (10, 3)-a and joining each to three others through tetrahedral tunnels. The arrangement of tunnels around a particular octahedron is shown

Fig. 17.2. The polyhedron O_{3t} (3, 8).

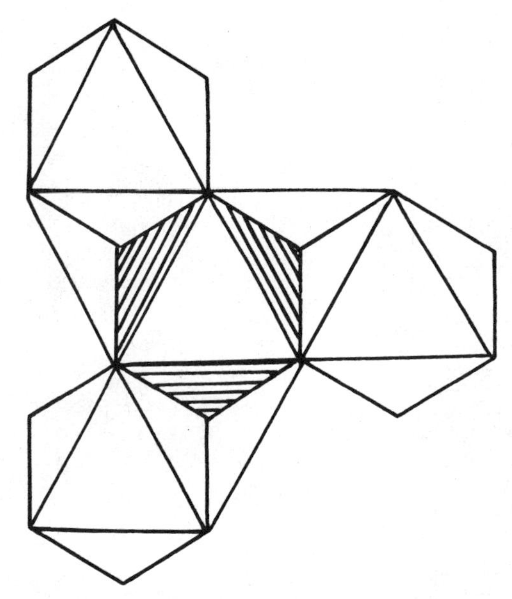

Fig. 17.3. Relation of tetrahedral tunnels to octahedra in O_{3t} (3, 8).

in Fig. 17.3, in which the shaded faces of the central octahedron are those shared with the tetrahedra. The three neighboring octahedra are coplanar with the central one, and this polyhedron may therefore be built by adding tetrahedra at the appropriate places to the edge-sharing octahedral structure of Fig. 7.3, which is based on the net (10, 3)-*a*. The tetrahedra in Fig. 17.2 form a 3D connected system in which each vertex is common to two tetrahedra; thus if built from AX_4 groups, the composition is AX_2 (silica-like structure). The A atoms are situated at the points of the $\left(\begin{matrix}3\\10\end{matrix}, 4\right)$ net derived by joining together the midpoints of the links of (10, 3)-*a* to form triangles around the nodes of the 3-connected net. The positions of the vertices of O_{3t} are three-quarters of those of cubic closest packing.

T_{4t} (3, 9)

In the polyhedron of Fig. 17.4, formed from tetrahedra at the nodes of the diamond net joined through (2-valent) octahedra along the links of that net, we can distinguish an edge-sharing octahedral structure in which each octahedron shares six edges with other octahedra. If built of AX_6 groups, the

Fig. 17.4. The polyhedron T_{4t} (3, 9).

Fig. 17.5. The polyhedron T'_{4t} (3, 9).

composition is AX_2. The X atoms are in the positions of cubic closest packing, and the structure is derived from the NaCl structure by removing the appropriate rows of A atoms. This is the idealized structure of atacamite, one of the polymorphs of $Cu_2(OH)_3Cl$.

T'_{4t} (3, 9)

If tetrahedra are added to T_{4t} as shown in Fig. 17.5, a different (3, 9) polyhedron is formed. Each of the added tetrahedra shares two faces with octahedra (2-valent tunnel as in O_{3t}). Although all these tetrahedra are equivalent, it is convenient to distinguish alternate tetrahedra along a strip of octahedra that share opposite edges in the way depicted in Fig. 17.6. Each octahedron shares one face with a tetrahedron of type b_1, one with b_2, and two with tetrahedra of type a, the latter being the tetrahedra of T_{4t}. The tetrahedra of type a are not visible in Figs. 17.4 or 17.5 because they are totally enclosed within the groups of four octahedra. Each set of added tetrahedra b_1 and b_2 forms a structure of the cristobalite type (idealized with –X– 180°), so that all the tetrahedra b_1 and b_2 together form two interpenetrating nets as in the cuprite structure. Note that T'_{4t} cannot be described as a 3D polyhedron in terms of octahedra and the tetrahedra b_1 and b_2 like O_{3t}, because an octahedron shares a face with only one of each. Since the octahedra and the tetrahedra a themselves form a 3D polyhedron T'_{4t} clearly has excessive tunnels and is therefore somewhat more complex than the other eight triangulated regular 3D polyhedra.

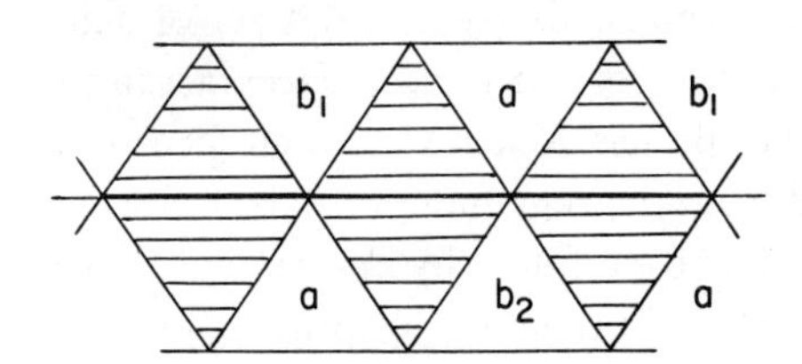

Fig. 17.6. Relation of tetrahedra to octahedra in T'_{4t}

O_{4t} (3, 8)

Octahedra at the nodes of the diamond net are joined through octahedra along the links of that net to form the polyhedron, of Fig. 17.7, which appears in Fig. 16.10 as a polyhedron with curved faces. The linking octahedra themselves form a vertex-sharing octahedral framework (of composition AX_3 if composed of AX_6 groups) which is the basis of the pyrochlore structure.

Fig. 17.7. The polyhedron O_{4t} (3, 8).

O_{6t} *(3, 10)*

As Fig. 17.8 indicates, octahedra at the points of a 6-connected net are connected through octahedral tunnels involving all but a pair of opposite faces. [This polyhedron is a different (3, 10) from that of Figs. 16.19*b* or 5.8.]

O_{8t} *(3, 12)*

The polyhedron of Fig. 17.9, built from octahedra joined through eight octahedral tunnels, is illustrated (with curved faces) in Fig. 16.20*b*.

I_{4t} *(3, 7)*

The polyhedron of Fig. 17.10 is formed from icosahedra linked to four others through octahedra. The regular icosahedron may be referred to orthogonal axes that pass through the midpoints of opposite edges and are axes of 2-fold symmetry. The faces fall into two groups, six pairs and a set of eight arranged octahedrally. Half of the latter, a group of four faces arranged tetrahedrally, are those involved in the tunnels. When four such faces of a regular icosahedron are distinguished from the remainder (e.g., by coloring) the icosahedron is enantiomorphic. It has been found by experiment that if all the icosahedra are D or L they do not join together to form a 3D polyhedron. It is necessary to join alternate D and L icosahedra to form the polyhedron of Fig. 17.10.

I_{6t} *(3, 8)*

The only polyhedron incorporating dodecahedral tunnels (Fig. 17.11) is built from "cubic icosahedra." The complementary polyhedron, an assembly of equal numbers of D and L snub cubes, is illustrated in Fig. 17.12 and, with curved faces, in Fig. 16.8*b*.

Fig. 17.8. The polyhedron O_{6t} (3, 10).

Fig. 17.9. The polyhedron O_{8t} (3, 12).

Fig. 17.10. The polyhedron I_{4t} (3, 7).

Fig. 17.11. The polyhedron I_{6t} (3, 8).

Fig. 17.12. The polyhedron (3, 8) built from snub cubes; the complementary polyhedron is I_{6t} (Fig. 17.11).

I_{8t} (3, 9)

The polyhedron of Fig. 17.13, built from icosahedra joined to eight others through octahedral tunnels, is illustrated with curved faces in Fig. 16.20*a*. The octahedra themselves form a connected system, each one sharing all its vertices with six others; therefore the system of octahedra is topologically similar to that in the ReO_3 and FeF_3 structures. The complementary polyhedron, based on the NbO net, is shown in Fig. 17.14.

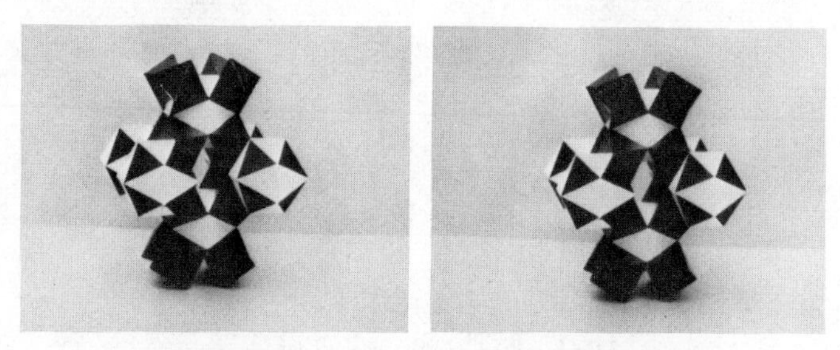

Fig. 17.13. The polyhedron I_{8t} (3, 9).

Fig. 17.14. The complement of I_{8t}.

Regular polyhedra $\{4, p\}$

As noted under the systematic description of curved-face polyhedra, all the (4, 5) polyhedra except one can be constructed with regular plane faces (squares). They comprise two 4-(coplanar tunnel) polyhedra and their 8-tunnel complements, a 5-tunnel polyhedron and its 8-tunnel complement, and two tetragonal 6-tunnel polyhedra that are identical to their complements—in all, eight polyhedra if we include the complements. We also describe two (4, 6) polyhedra, one identical with and the other different from its complement.

(4, 5)-4t, types a and b of Table 16.2

There are two (4, 5) regular 3D polyhedra based on the NbO net. The first (Fig. 17.15) is a partial ($\frac{9}{32}$) space-filling by cubes that corresponds to the curved-face polyhedron of Fig. 16.22*a*. The second (Fig. 17.16) is not built from cubes but is the complement of the polyhedron (Fig. 17.20) constructed from truncated octahedra and hexagonal prisms.

(4, 5)-5t, type c of Table 16.2

This strictly (3 + 2)-tunnel polyhedron appears in projection in Fig. 17.17*a* and as a stereo-pair in Fig. 17.18; it is based on the 5-connected 3D net of Fig. 15.1*e*. The polyhedron is formed from columns of hexagonal prisms sharing opposite hexagonal faces (which form the two larger tunnels), with three tunnels emerging at the same level from alternate square faces of alternate prisms in each column. The complementary polyhedron is based on columns of dodecagonal prisms. The three tunnels from a given unit must be at the same height for the reason stated under (4, 6)-5*t* and complementary (4, 6)-8*t*.

(4, 5)-6t, types e and f of Table 16.2

Figure 17.19*a* shows a unit of the 6-tunnel, or strictly (4 + 2)-tunnel polyhedron based on the *P* lattice and formed from columns of octagonal prisms. All four horizontal tunnels may be at the same height or they may be at heights 0 and $\frac{1}{2}$ (Fig. 17.19*b*), as shown in the projections of Fig. 17.17*b* and *c*. This type of variant is possible only for this tetragonal polyhedron and not for the trigonal (4, 5) polyhedron of Fig. 17.17*a* or for the hexagonal (4, 6) polyhedron of Fig. 17.17*d*. The complements of both the polyhedra of Fig. 17.19

Fig. 17.15. The regular 3D polyhedron (4, 5)-4*t* (type *a* of Table 16.2).

Fig. 17.16. The regular 3D polyhedron (4, 5)-4t (type b).

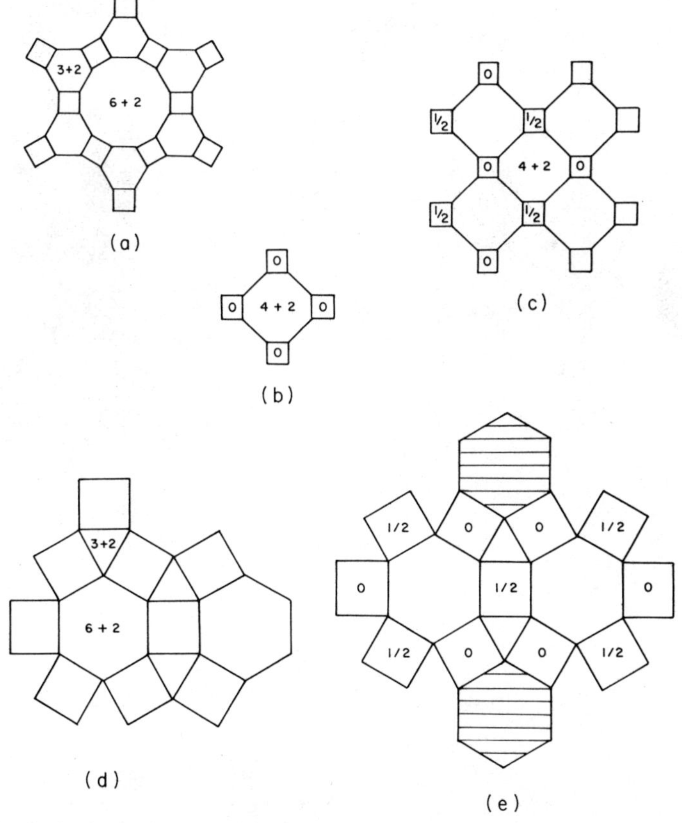

Fig. 17.17. Projections of regular 3D polyhedra. (*a*) (4, 5)-5t and complementary (4, 5)-8t; (*b*), (*c*) (4, 5)-(4 + 2)t and variant; (*d*) (4, 6)-5t and complement; (*e*) see text.

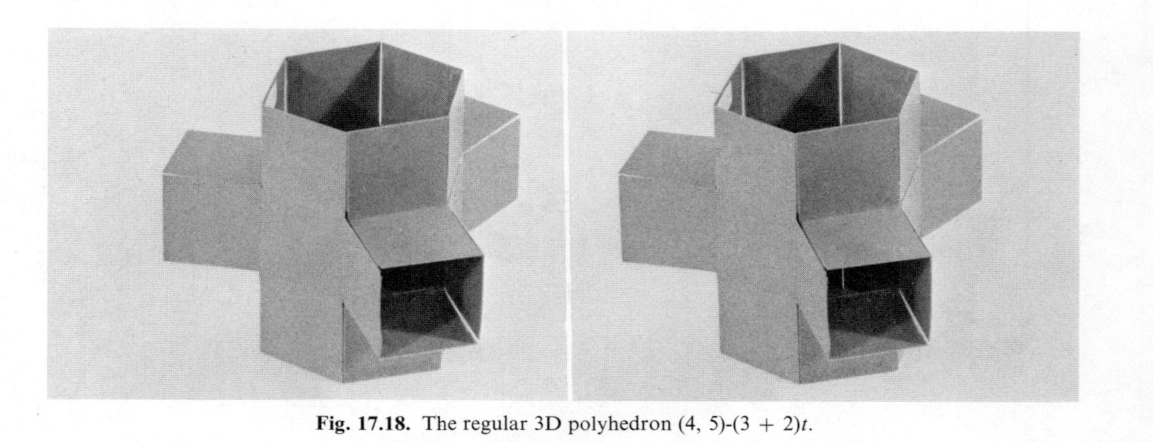

Fig. 17.18. The regular 3D polyhedron (4, 5)-(3 + 2)*t*.

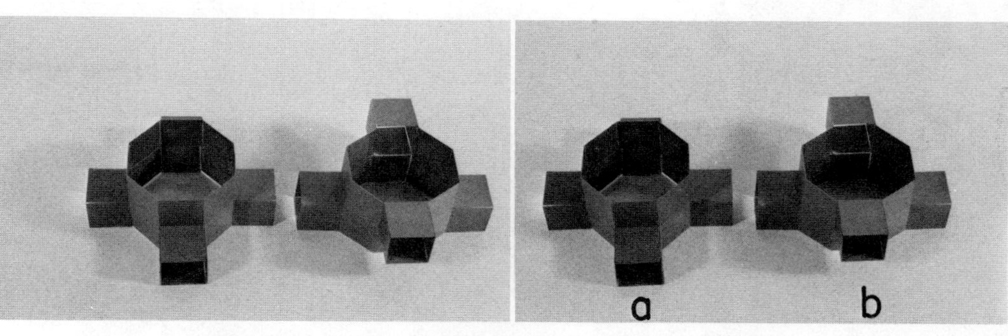

Fig. 17.19. The regular 3D polyhedra (4, 5)-6*t* (types *e* and *f*).

are the same as the polyhedra themselves; they are the only members of the (4, 5) class of which this is true. (If the tunnels are put at heights 0, $\frac{1}{4}$, $\frac{1}{2}$, and $\frac{3}{4}$, to form helical arrays around the columns of octagonal prisms, there is formed a "Catalan" type of polyhedron with 4^4 and 4^5 vertices.)

The remaining regular (4, 5) polyhedra are the complements of the first three.

(4, 5)-8t, type g of Table 16.2

The complement of the polyhedron of Fig. 17.18 is strictly a (6 + 2)-tunnel polyhedron, with two large vertical tunnels and six horizontal tunnels emerging from alternate square faces of dodecagonal prisms.

(4, 5)-8t, types h and i of Table 16.2

These polyhedra are based on the *I* lattice and, unlike the preceding polyhedron, have eight equivalent tunnels. One repeat unit of type (*h*) contains 24 points and can be described as a block of 27 cubes, of which the eight at the corners are shared with adjacent units. It is seen at the center of Fig. 17.15. The polyhedron of type *i* is the complement of that of Fig. 17.16 and is illustrated in Fig. 17.20.

Fig. 17.20. The regular 3D polyhedron (4, 5)-8*t*, complementary to that of Fig. 17.16.

(4, 6)-6t

This is a half-space-filling by cubes (Fig. 17.21*a*) and is Coxeter's skew polyhedron {4, 6/4}.

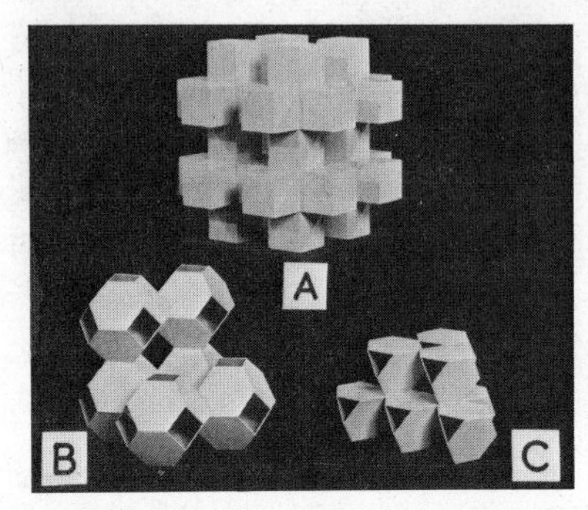

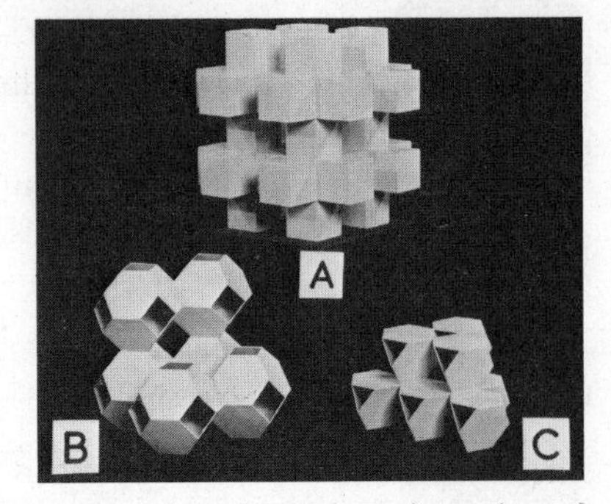

Fig. 17.21. The 3D regular skew polyhedra of Coxeter. (*a*) {4, 6/4}; (*b*) {6, 4/4}; (*c*) {6, 6/3}.

(4, 6)-5t and complementary (4, 6)-8t

These polyhedra are shown in projection in Fig. 17.17*d* and as a stereo-pair in Fig. 17.22, which includes one complete 8-tunnel unit and one 5-tunnel unit of the complementary polyhedron. The 8-tunnel polyhedron is formed from hexagonal prisms stacked with their hexagonal faces shared, six horizontal tunnels emerging at the same level from all square faces of alternate prisms in any vertical column. Of the eight tunnels from each unit, six therefore correspond to square faces and two to hexagonal faces of the prism. The variant with the horizontal tunnels alternately at heights 0 and $\frac{1}{2}$ around each column of prisms is not admissible because it implies that all units are not identical. Placing tunnels alternately at heights 0 and $\frac{1}{2}$ from two adjacent units gives pairs of adjacent tunnels at the *same* height from the units distinguished by shading in Fig. 17.17*e*.

Fig. 17.22. The 3D polyhedra (4, 6)-5*t* and (4, 6)-8*t*.

Regular polyhedra $\{6, p\}$

(6, 4)-6t

This is the half-space-filling array of truncated octahedra illustrated in Fig. 17.21*b* and is Coxeter's skew polyhedron $\{6, 4/4\}$.

The third of Coxeter's skew polyhedra, namely $\{6, 6/3\}$, is a half-space-filling by equal numbers of tetrahedra and truncated tetrahedra (Fig. 17.21*c*). It is not admissible as a polyhedron (6, 6) of the type we are considering because it is built of units of two kinds.

18

Complementary three-dimensional polyhedra

We define a complementary polyhedron as that space not enclosed by the surface that encloses the polyhedron, subject to the requirement that it has the same properties as the polyhedron itself, namely: (*a*) it is based on a periodic 3D net in which all points are p-connected and topologically equivalent, and (*b*) it is formed by repetition of identical polyhedral units at the nodes of such a net, these units being joined directly by sharing faces or through tunnels which connect two units.

It follows from *a* that a polyhedron and its complement are based on a pair of complementary interpenetrating nets, of which five pairs were listed in Table 11.2. It appears that all known examples of polyhedra (n, p) and their complements are based on one of the following four pairs of nets, 1, 2, 3, and 4 of Table 11.2:

Nets, Table 11.2	p	Polyhedron	p	Complement
1	6	*P* net	6	*P* net
2	4	Diamond	4	Diamond
3	4	NbO net	8	*I* net
4	5	Fig. 15.1*e*	8	Fig. 11.13

In pair 1 all six tunnels may be the same type (cubic symmetry) or they may be four of one kind and two of another (tetragonal symmetry). The summarizing Table 19.1 distinguishes between these two types of polyhedron based on a P lattice as having 6 or $4 + 2$ tunnels. It is seen that *some* of the former have polyhedron and complement identical, whereas *all* the latter are half-space-filling. The only example of pair 2 is the (5, 5) polyhedron based on the "compressed-diamond" net, which is identical with its complement. All four or all eight of the tunnels are similar in pair 3, and here the polyhedron and complement are not identical. In pair 4 the tunnels are of two kinds, $(3 + 2)$ trigonal bipyramidal, and $(6 + 2)$ hexagonal bipyramidal; polyhedron and complement are not identical.

It should be noted further that the complement of an I type polyhedron is of the NbO type (built of identical units) only if the 8-tunnel unit has cubic symmetry. The 8-tunnel unit of the (3, 7) polyhedron of Fig. 16.7 has only *mmm* symmetry, and its complement is built of units of two kinds. The 8-tunnel units occupy the positions (000) and $(\frac{1}{2}\frac{1}{2}\frac{1}{2})$ in the unit cell of Fig. 18.1. The horizontal symmetry planes are at heights 0 and $c/2$. The vertical symmetry planes are parallel to (110) and $(\bar{1}10)$, and therefore the units A and B situated at the points corresponding to the NbO net have point symmetries *mmm* and *m*, respectively. The NbO type polyhedron is composed of two A and four B units per unit cell.

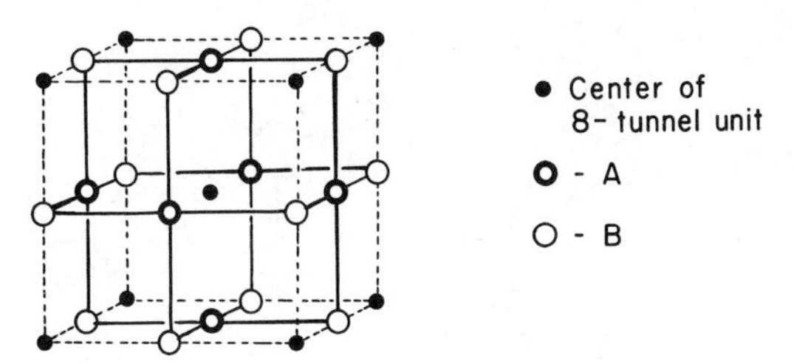

Fig. 18.1. The two kinds of repeat unit (A and B) in the 4-tunnel complement of an 8-tunnel polyhedron having noncubic symmetry.

The value of p for the underlying net is, of course, the number (t) of tunnels for the polyhedron. For a given family of 3D polyhedra $\{n, p\}$ (with different values of t), the values of Z_t are proportional to $(t - 2)$; for example, $Z(8t) = 2Z(5t) = 3Z(4t)$. Therefore if the complement of a 3D polyhedron is based on a unit having a number of tunnels *different from* that of the polyhedron itself, the same surface can be described in terms of different (topological) repeat units with different values of Z_t:

Relation between polyhedron and complement	Relation between values of Z_t
NbO net: *I* net	$3Z_t(4t) = Z_t(8t)$
(3 + 2) net: (6 + 2) net	$2Z_t(5t) = Z_t(8t)$

The number of points (Z_c) in the crystallographic repeat unit is necessarily the same for either description, since the surface is the same for polyhedron and complement; the arbitrariness in choice of Z_t may therefore be avoided by giving Z_c rather than Z_t. [The *numbers* of repeat units of polyhedron and complement *in a given volume* are in the inverse ratios, which are the same as the ratios of numbers of points in a given volume of the underlying nets. The unit cell of Fig. 18.1 contains two points of the *I* net but six of the NbO net, and the unit cell of the 5-connected net 3 of Table 11.2 contains twice as many points as that of the 8-connected net (Fig. 18.2).]

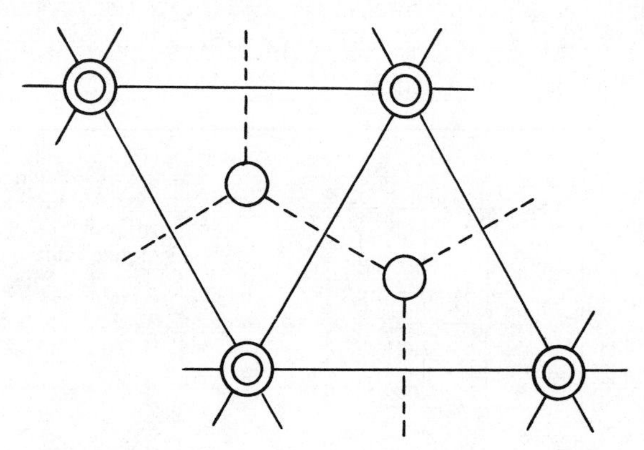

Fig. 18.2. Projection of unit cell of (3 + 2)- or (6 + 2)-connected nets.

There is clearly the same choice of Z_t for nets $(n, 3)$, for example, which are duals of surface tessellations $(3, p)$. This difficulty does not arise for 6-tunnel polyhedra and dual nets based on the *P* net (because the complement also is based on the *P* net), though of course Z_c may still be a multiple of Z_t if there are differently oriented units in the unit cell. For example, for the (3, 10)-6*t* polyhedron of Fig. 5.8, $Z_t = 6$ but $Z_c = 48$, and for the dual net (10, 3)-*g*,

$Z_t = 20$ but $Z_c = 160$. (The values of Z_c are equal to $8Z_t$ because (i) there are four different orientations of the basic unit and (ii) the cell is body-centered.)

The absence of pairs of complementary triangulated polyhedra based on the diamond net calls for comment. Apart from the (5, 5) polyhedron based on the "compressed-diamond" net the only known 3D polyhedra based on the diamond net are triangulated ones (Table 19.1),* but none of these polyhedra (3, 7), (3, 8), or (3, 9) has a complement that is a 3D polyhedron. In the derivation of the modified Euler relation (p. 197) it is supposed that a polyhedral repeat unit can be carved out of the 3D polyhedron (Fig. 15.3), from which it follows that the complementary polyhedron must also be based on some kind of polyhedral repeat unit. Now the complement to a polyhedron based on the diamond net would also be based on this (4-tunnel) net (see Table 11.2) therefore would have the same value of Z as the polyhedron itself. It is simplest to consider first a particular example, such as the 4-tunnel polyhedron (3, 8) of Fig. 16.10. The large channels would form the tunnels emerging from a unit of the complement. The smallest circuit around one of these contains six points, as compared with three points around a tunnel of the polyhedron itself. The polyhedral 4-tunnel repeat unit of the complement would therefore have to be a polyhedron with six vertices [$Z = 6$ for a (3, 8)-4t polyhedron] and four 6-gon faces; there is no such polyhedron.

* The half-space-fillings by tetrahedra and truncated tetrahedra [Coxeter's skew polyhedron {6, 6/3}, Fig. 17.21c], and by rhombic dodecahedra (Fig. 18.3), do not conform to our rather restrictive definition of a 3D polyhedron. The former is built of two kinds of unit, each with four tetrahedral neighbors of the other kind; the present account is confined to polyhedra built of identical units. The latter, which is based on the "extended-diamond" net, is excluded because it is not a 3D polyhedron of the type (n, p) but has nonequivalent vertices; that is, it is of the Catalan type $\left(4, \begin{matrix}4\\8\end{matrix}\right)$.

Fig. 18.3. Half-space-filling by rhombic dodecahedra.

The corresponding figures for other polyhedra based on the diamond net are as follows:

	Polyhedron		
	(3, 7)	(3, 8)	(3, 9)
Z	12	6	4
Minimum circuit around tunnel of complement	12	6	6
Minimum circuit around tunnel of polyhedron	3	3	3

Because the required polyhedra do not exist; it follows that there are no complements to triangulated polyhedra based on the diamond (or "pseudo-diamond") net that would conform to requirements *a* and *b* stated earlier.

The 3D polyhedra we have been discussing are of the type (n, p); that is, they are the homologues of the Platonic solids. We have mentioned that the half-space-filling by rhombic dodecahedra is of the Catalan type, and we noted in the previous section that space-filling arrangements of polyhedra provide examples of Archimedean-type 3D polyhedra. From a space-filling arrangement of cubes, truncated octahedra (*ot*), and truncated cuboctahedra (*cot*), we may separate the cubes and *ot*'s as forming a 3D polyhedron (Fig. 18.4) based on the *P* net (point symbol $4^2 6^2$, complete symbol as 4-connected net, $4^3 6^2 8$). The complementary polyhedron is built from *cot*'s

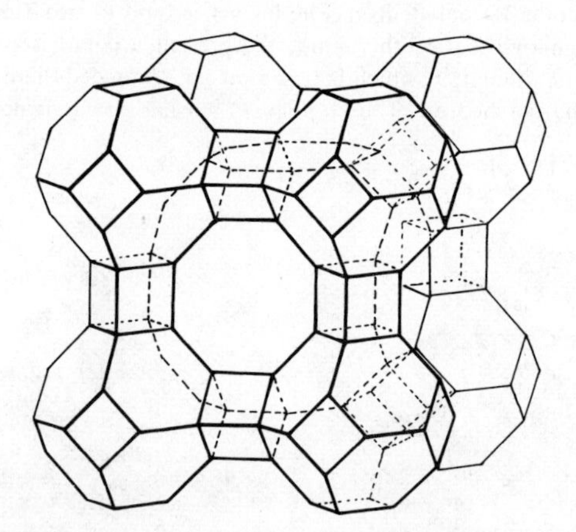

Fig. 18.4. Open packing of truncated octahedra and cubes.

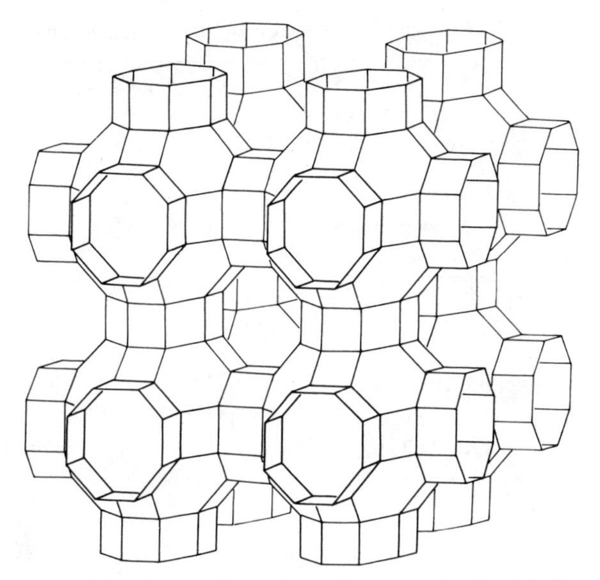

Fig. 18.5. Half-space-filling by equal numbers of truncated cuboctahedra and octagonal prisms.

sharing their 8-gon faces. The space-filling arrangement of *cot*'s and octagonal prisms (p_8) illustrates two semiregular 3D polyhedra $4^2 8^2$ and $4^3 6$ and also the breakdown of a polyhedral space-filling into two different pairs of complementary polyhedra. The positions of the two kinds of polyhedra are those of the shaded circles (*cot*) and open circles (p_8) of Fig. 11.2*c*. Each *cot* is in contact with eight others (sharing 6-gon faces); thus there is a continuous b.c. system of *cot*'s. The octagonal prisms also form a continuous system by sharing alternate 4-gon faces to form a 4-tunnel 3D polyhedron ($4^2 8^2$) based on the NbO net, which is complementary to the polyhedron formed from the *cot*'s. Alternatively, one half of the *cot*'s and one half of the p_8's form a 6-tunnel 3D polyhedron $4^3 6$ based on the primitive lattice that is identical with its complement (Fig. 18.5):

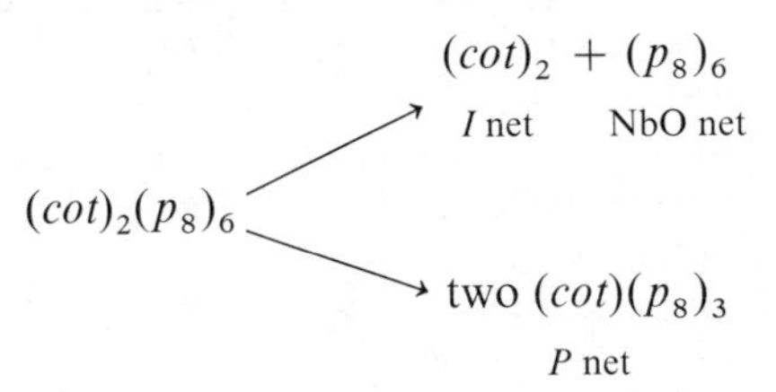

Table 18.1 Some 3D polyhedra and their complements

Basic nets	Type of polyhedron: Regular plane faces	Curved faces
Platonic		
P lattice	Skew polyhedron {4, 6/4} (4^6)* (half-space-filling by cubes) Skew polyhedron {6, 4/4} (6^4)* (half-space-filling by truncated octahedra)	(3, 8) Fig. 16.8*c** (3, 9) Fig. 16.19*a** (3, 10) Fig. 16.19*b**
	(3, 8) Figs. 17.11 and 17.12	(3, 8) Fig. 16.15
NbO and *I*	(3, 9) Figs. 17.13 and 17.14	
	(4, 5)-4*t* Figs. 17.16 and 17.20	(4, 5) Fig. 16.22
(3 + 2) and (6 + 2)	(4, 5)-5*t* Fig. 17.22	(3, 8)-5*t* Fig. 16.16
Archimedean		
P lattice	$4^3 6$ Fig. 18.5* (half-space-filling by *cot* + p_8) $4^2 6^2$ Fig. 18.4 (partial space-filling by cubes + *ot* or by *cot*)	
NbO and *I*	$4^2 8^2$ (partial space-filling by p_8 or by *cot*)	
Catalan		
Diamond	$\left(4, \begin{matrix}4\\8\end{matrix}\right)$ Fig. 18.3* (half-space-filling by rhombic dodecahedra)	
Polyhedron built of units of two kinds		
Diamond	Skew polyhedron {6, 6/3} (6^6)* (half-space-filling by tetrahedra and truncated tetrahedra)	

* All 3D polyhedra that are identical with their complements are half-space-filling. The volume of a given 3D polyhedron with regular *plane* faces is determined by the edge length only. This is also true of all curved-face polyhedra constructed with *equilateral* faces. The various curvatures (hence the volumes of the polyhedra) are not variable for a given tessellation on a given surface. Without the condition of equilateral faces, the volume would be infinitely variable, since the surface of a polyhedron arises by inflating the links of a 3D net.

Note that the point symbols in the preceding paragraph and for these polyhedra in Table 18.1 refer to the surface of the 3D polyhedron. For example, the complete point symbol of the half-space filling of Fig. 18.5 regarded as a 4-connected net is $4^3 6^3$ (Table 10.1). Some information about pairs of complementary polyhedra is summarized in Table 18.1 and a number of examples are illustrated as indicated in the table.

19

Summary of three-dimensional polyhedra

On the surfaces (of varying curvature) formed by inflating the links of 3D nets, various tessellations may be inscribed which we call 3D polyhedra. We have confined our attention for the most part to tessellations (n, p), that are solutions of the equation

$$\sum f_n[4 - (2 - n)(2 - p)] = 2p(2 - t)$$

for $t \geqslant 3$, where t is the number of tunnels emerging from each repeat unit; this number is the connectedness of the basic net. (The solutions corresponding to $t = 0$, 1, and 2 are also described, and examples are noted of polyhedra based on 2D nets.)

The 3D polyhedra have the following properties.

1. Instead of the single solutions (n, p) for finite polyhedra (the Platonic solids), there is a family of 3D polyhedra $\{n, p\}$ for given values of n and p, corresponding to different values of t—that is, based on different 3D nets. For a given value of t, the number of points in the repeat unit (Z) decreases as n and/or p increase, as shown for 6-tunnel polyhedra in Table 16.1. The number of 3D polyhedra is therefore limited, though it is not known how many are actually realizable.
2. In some cases topologically different tessellations having the same values of n and p can be inscribed on surfaces derived from the same 3D net. Examples are included in the summarizing Table 19.1.

Table 19.1 Curved-surface 3D polyhedra

Number of tunnels, t	3D net	(3, p)					(4, p)			(5, p)
3	(10, 3)-a	(3, 7)*	(3, 8)							
	(10, 3)-b	(3, 7)†	(3, 8)							
4	Diamond	(3, 7)	(3, 8) (3, 8)*	(3, 9)						(5, 5)‡§
	NbO‖		(3, 8)	(3, 9)		(3, 12)	(4, 5) (two)			
3 + 2	(3 + 2)-net#		(3, 8)				(4, 5)			
6	P lattice	(3, 7)*	(3, 8)		(3, 10)§		(4, 5)§	(4, 6)§		
4 + 2			(3, 8)§	(3, 9)§			(4, 5)§ (two)	(4, 6)§		
8	I lattice	(3, 7)	(3, 8) (two)	(3, 9)		(3, 12)	(4, 5) (two)		(4, 7)	
6 + 2	(6 + 2)-net		(3, 8)				(4, 5)			
[12	See page 221		(3, 8)]							

* Cannot be constructed with equal links (see pp. 211, 213).
† See page 211.
‡ "Compressed-diamond" net.
§ Half-space-filling.
‖ Entries are the same as for the complementary I net except where 8-tunnel unit does not have cubic symmetry (p. 251).
Entries are the same as for the complementary (6 + 2)-net.

3. On a particular type of surface (derived from a given 3D net) only the tessellations (n, p) listed in Table 19.1 have been found; further study may reveal additional 3D polyhedra.

4. As in the case of finite polyhedra and plane nets, there are dual tesselations on the same surface. For example, the members of a pair such as (4, 5) and (5, 4), Figs. 19.1 and 19.2, are related in the same way as the

Fig. 19.1. The 3D polyhedron (4, 5)-4t.

Fig. 19.2. The 3D polyhedron (5, 4)-4*t*.

icosahedron (3, 5) and the dodecahedron (5, 3). Like (3, 3) and (4, 4), (5, 5) is self-dual.

5. Provided there are no circuits around the tunnels smaller than n-gons, the links (edges) of a 3D polyhedron (n, p) may be described as a 3D net. Nets that satisfy the criteria for uniformity set out on page 14, namely,

p:	3	4	5	6
point symbol:	n^3	n^6	n^9	n^{12}

include numerous systems (n, 3), the only known member of the {5, 5} family, which is 5^9, and the (4, 6)-6 tunnel polyhedron, which is also the net 4^{12}; see page 225.

6. The dual relation between 3D polyhedra and uniform nets is limited as described in Chapter 20.

7. Certain polyhedra (n, p) can be constructed with plane equilateral faces; they are the 3D homologues of the Platonic solids. There are also homologues of the Archimedean and Catalan solids.

8. The space that is not occupied by a 3D polyhedron is described as the *complementary polyhedron*. In certain cases the polyhedron and its complement are identical; that is, the surface divides space into two equal parts (Table 18.1).

20

Relation of three-dimensional polyhedra to three-dimensional nets: Dual relations between them

We now return briefly to a subject discussed in the introductory section. Along the top row of Table 1.1 there are triangulated 3D polyhedra, tessellations of triangles on well-defined surfaces formed by inflating the links of 3D nets. Down the left-hand column there are uniform 3-connected nets, and these are of two distinct types: (*a*) duals of certain uniform polyhedra [e.g., the uniform nets (8, 3)-*e*, *f*, etc.], and (*b*) other uniform nets that do not at all have the characteristics of 3D polyhedra. The polyhedral surfaces of nets of type *a* become less obvious as *n* increases, and it might seem as if there could be a gradual transition across the table from 3D polyhedra to uniform 3D nets. What, in fact, is the relation between these two kinds of 3D system? A related matter is that a dual relationship exists between the pairs of Platonic solids (3, 4), (4, 3) and (3, 5), (5, 3) and between the plane nets (3, 6) and (6, 3), which are symmetrically placed in relation to the diagonal line in Table 1.1. Our second question is: does this dual relationship exist more generally between the 3D systems of Table 1.1?

Our conclusions regarding the first question are as follows.

1. A given 3D polyhedron (n, p) is not necessarily a uniform net; it *may* be a uniform net if there are no circuits around the tunnels smaller than n-gons.
2. A given uniform net (n, p) is not necessarily a 3D polyhedron. The property of being a 3D polyhedron is an *additional* property of certain uniform nets. The first requirement is that the value of Z_t for the net equal or exceed a certain minimum value. In a given family of polyhedra $\{n, p\}$ the value of Z_t for the 3-tunnel unit sets a lower limit below which a uniform net cannot be a 3D polyhedron; but there are further limitations on the value of Z_t, as well. We discuss here the polyhedra $(n, 3)$. The values of Z_t for polyhedra $(n, 3)$ are listed in Table 20.1; for higher numbers of tunnels the value of Z_t is a multiple of the value for the 3-tunnel polyhedron. Evidently any uniform (3-connected) net (7, 3) having Z_t less than 14 cannot be a 3D polyhedron; moreover, if any (7, 3) net is also to be a 3D polyhedron, the value of Z_t must be a multiple of 14. For reasons given in Table 20.1 the realistic minimum value of Z_t is in some cases probably, and in other cases certainly, higher than the value for the 3-tunnel polyhedron. For example, no 3-connected net having Z_t less than 10 (or more probably 20) is the dual of a 3D (3, 10) polyhedron. The first (10, 3) net known to be such a dual is (10, 3)-g, for which

Table 20.1 Permissible values and probable minimum values (boldface type) of Z_t for 3D polyhedra $(n, 3)$

	Z						
Polyhedron	$3t$	$4t$	$5t$	$6t$	$8t$	$12t$	Notes
(7, 3)	**14**	28	42	56	84	140	
(8, 3)	**8**	16	24	32	48	80	
(9, 3)	6	**12**	18	24	36	60	For (3, 9)-$3t$, $Z = 2$; lowest known value of t is 4
(10, 3)	5	10	15	**20**	30	50	*a.* Z must be even for a 3-connected net *b.* Lowest known value of t is 6
(12, 3)	4	8	12	16	**24**	40	Lowest known value of t is 8

$Z_t = 20$. Similarly the nets (7, 3)-*a* and (7, 3)-*b*, for both of which $Z = 12$, are not duals of 3D polyhedra; but (7, 3)-*c* and (7, 3)-*d*, with Z_t equal to 14 and 84, respectively, are both duals of 3D polyhedra. The low value of the minimum value of Z_t for duals of polyhedra (3, 8) may be in part responsible for the large number of uniform (8, 3) nets, many of which (see Table 5.3) are in fact duals of, therefore are themselves, 3D polyhedra.

We can certainly say that a system $(n, 3)$ cannot be a 3D polyhedron unless Z_t is equal to the appropriate minimum value given in boldface type in Table 20.1 or one of the higher values shown (multiple of the value for the 3-tunnel polyhedron). It does not, however, appear that satisfying this requirement is a sufficient condition, for there are some uniform nets that satisfy the Z_t requirement but apparently are not 3D polyhedra. Nets (7, 3), (10, 3), and (12, 3) present no problems, but the following uniform nets apparently are not duals of 3D polyhedra, therefore are not themselves 3D polyhedra:

nets (8, 3):	*c*,	*d*,	*g*,	*h*,	*l*
Z:	8	8	16	16	24
nets (9, 3):	*a*	*c*			
Z:	12	24			

Further study of this point would be desirable.

We have now in effect answered our second question. Clearly every 3D polyhedron (n, p) has a dual that is a tessellation on the same surface (with n and p interchanged); systems such as (5, 5), like (3, 3) and (4, 4), are self-duals. However neither the polyhedron (n, p) nor its dual (which is also a 3D polyhedron) is necessarily a uniform net. A dual relationship between a 3D polyhedron and a uniform 3D net exists only in a limited number of cases, namely, when the uniform net is also a 3D polyhedron.

Index of Chemical Formulae

Subject Index